Lorenzo Massimo Polgar, Machiel van Essen, Andrea Pucci, France

Smart Rubbers

Also of interest

Magneto-Active Polymers
Fabrication, characterisation, modelling and simulation
at the micro- and macro-scale
Pelteret, Steinmann, 2019
ISBN 978-3-11-041951-1, e-ISBN 978-3-11-041857-6

Twin Polymerization
New Strategy for Hybrid Materials Synthesis
Spange, Mehring, 2019
ISBN 978-3-11-050067-7, e-ISBN 978-3-11-049936-0

Shape Memory Polymers
Kalita, 2018
ISBN 978-3-11-056932-2, e-ISBN 978-3-11-057017-5

Polymer Engineering
Tylkowski, Wieszczycka, Jastrzab (Eds.), 2017
ISBN 978-3-11-046828-1, e-ISBN 978-3-11-046974-5

e-Polymers.
Editor-in-Chief: Seema Agarwal
ISSN 2197-4586
e-ISSN 1618-7229

Lorenzo Massimo Polgar, Machiel van Essen,
Andrea Pucci, Francesco Picchioni

Smart Rubbers

Synthesis and Applications

2nd Edition

DE GRUYTER

Author
Dr. Lorenzo Massimo Polgar
NXP Semiconductors Netherlands B.V.
High Tech Campus 60
5656 AG Eindhoven
Netherlands

Machiel van Essen
Technical University of Eindhoven
Dept. of Chemical Engineering & Chemistry
PO Box 513
5600 MB Eindhoven
Netherlands

Prof. Andrea Pucci
Università di Pisa
Dipartimento di Chimica e Chimica Industriale
Via Giuseppe Moruzzi 13
56124 Pisa
Italy

Prof. Dr. Francesco Picchioni
University of Groningen
Faculty of Science and Engineering
Product Technology
Nijenborgh 4
9747 AG Groningen
Netherlands

ISBN 978-3-11-063892-9
e-ISBN (E-BOOK) 978-3-11-063901-8
e-ISBN (EPUB) 978-3-11-063931-5

Library of Congress Control Number: 2018966601

Bibliographic information published by the Deutsche Nationalbibliothek
The Deutsche Nationalbibliothek lists this publication in the Deutsche Nationalbibliografie;
detailed bibliographic data are available on the Internet at http://dnb.dnb.de.

© 2019 Walter de Gruyter GmbH, Berlin/Boston
Typesetting: Integra Software Services Pvt. Ltd.
Printing and binding: CPI books GmbH, Leck
Cover image: Photo ephemera / Moment / Getty Images

www.degruyter.com

Preface

'Smart rubbers' are defined as elastomeric materials that respond to external stimuli through a macroscopic output in which the energy of the stimulus is transduced appropriately as a function of external interference.

The bulk of these materials generally consist of elastomers or materials with rubbery behaviour and properties, but display a smart response. A 'rubber' is typically a polymer that completely recovers its original shape upon deformation as a result of physical or chemical crosslinking of the macromolecular network. The term 'smart' implies that the material displays a response to an external stimulus and that this occurs in a controlled manner.

Research into smart rubbers has increased drastically over the last few decades, predominantly due to the growing demand of, and the need for, improved materials for new applications. Incentives for these demands are generally based on societal aspects such as economics (i.e., cost reduction) and sustainability. Thus, smart rubbers are not limited only to scientific aspects but also intersect with societal relevance. An overview of how the research field of smart rubbers is divided and connected into several components is shown in Figure P1.1.

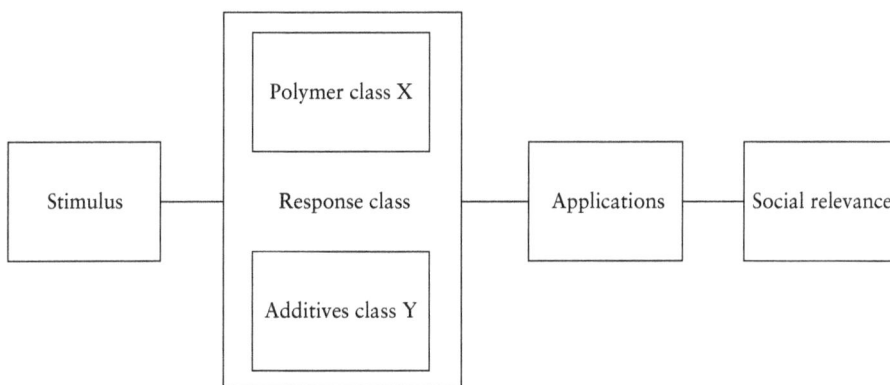

Figure P1.1: Schematic overview of the relationship between the various key variables that should be considered in the design of smart rubbers.

The research field of smart rubbers covers multiple components. Providing an update of smart rubbers for each component in one book is not possible. Hence, the authors have taken 'snapshots' of several contributions considered suitable to provide paradigmatic information about the main topic.

First, an update is provided on the sustainable design of smart rubbers (Chapter 1), which is a hot topic that has an evident high societal relevance. Subsequently, sensing rubbers (Chapter 2) represent a typical application of smart rubbers, and are discussed in the context of academic and industrial developments. Optically active

https://doi.org/10.1515/9783110639018-201

elastomers (Chapter 3) are addressed as another illustrative example of a different response category. Lastly, the combination of components required to formulate a smart rubber in a specific response class of actuating elastomers are discussed in more detail. This discussion is subdivided into shape memory (Chapter 4), magneto-rheological elastomers (Chapter 5), and dielectric elastomers (Chapter 6). Different stimuli and applications are entwined throughout all chapters.

In this way, the authors provide an update of smart rubbers that is relevant, interesting and understandable for industry, scientists, experts, and students.

Contents

1 Sustainability in the design of rubber materials

Sustainability becomes an ever more important and unavoidable topic when designing materials or chemical products. This is also the reason for the vast increase in research output on this topic and the number of scientific articles having 'sustainability' as a major keyword (Figure 1.1). Sustainable materials can be defined broadly as materials that can be produced and reused indefinitely without affecting the human–ecosystem equilibrium. Three major issues should be taken into consideration for the development of a sustainable material: the production process, carbon footprint and purpose of the material at the end of its product life.

First, the process to develop and produce the material should be sustainable. Hence, an energy-neutral process is required and all solvents and utility streams must ultimately be recycled without the production of any waste. A good example is the recently developed process to extrude rubbers using supercritical carbon dioxide ($scCO_2$) [1]. Besides being a good and green solvent, $scCO_2$ improves the dispersion of materials such as fillers or other polymers in the rubber matrix [1, 2]. After the process, the harmless solvent is easily fed back into the environment by releasing the pressure. The focus of this book is materials (smart rubbers) so, rather than smart processing, the carbon footprint of a material and its recyclability will be discussed in more detail because they are more related to the material itself. As both components are important, they cannot always be distinguished (Figure 1.2).

Currently, most rubber products consist of elastomers that originate from the lower-left quadrant. These elastomers are usually oil-based and non-recyclable due to the presence of irreversible crosslinks. A significant amount of the rubber products that we use ($\approx$40% of all rubber used [3, 4]) still contain a lot of natural rubber (NR), which is a natural polymeric compound produced from the latex of *Hevea brasiliensis* (which mainly contains *cis*-polyisoprene). This amount is ever decreasing as NR is being replaced by synthetic rubbers that have been developed to meet the highly demanding requirements for specific applications. A good example of such a specific application that requires very specific material is sealants. The neoprene group of synthetic rubbers that is generally used for such applications is very stable with respect to NR and is, therefore, used for multiple applications such as wetsuits, laptop sleeves and durable medical devices. The low electrical conductivity of these neoprene rubbers also makes them useful as an insulator in electrical wiring. NR is a poor candidate for such applications because the high number of unsaturated carbon–carbon double bonds makes the elastomers too reactive with other chemicals to be widely used for such applications. Recent developments using biomass as a feedstock for base chemicals have led to the replacement of some synthetic rubbers by bio-based alternatives. Two examples of such bio-based synthetic rubbers are discussed in Section 1.1. Even when using NR or bio-based elastomers in a rubber product, the final products are not completely bio-based due to large

https://doi.org/10.1515/9783110639018-001

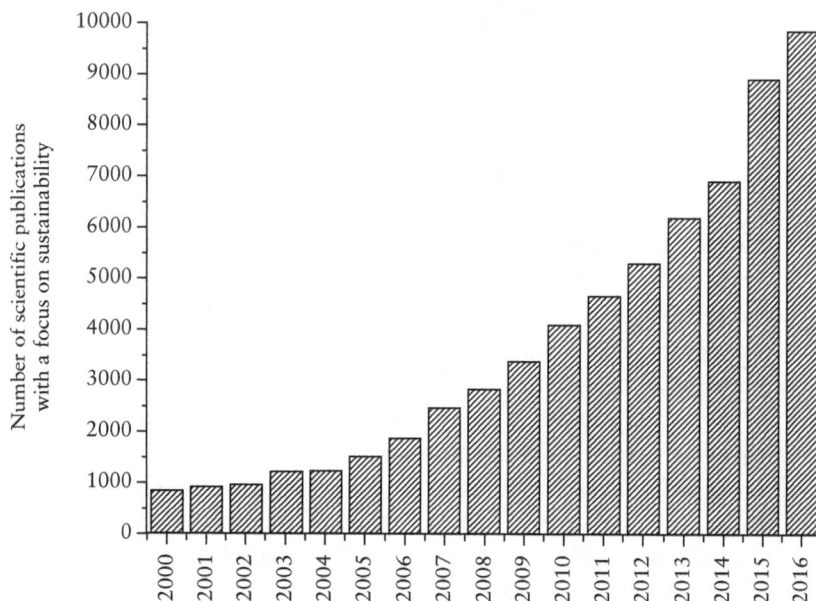

Figure 1.1: The exponential increase in scientific article that focus on sustainability over the last decades as generated from the 'web of knowledge'.

Figure 1.2: The environmental axis. The bio-based character and recyclability of materials as individual components of sustainability. Reproduced with permission from Dr. Martijn Beljaars.

amounts of additives such as (reinforcing) fillers or lubricants. Some more 'green' and sustainable alternatives for these oil-based fillers are also discussed.

Nevertheless, even if the process and materials used to produce rubbers are completely 'green', the materials are spent after some time, resulting in the accumulation of rubber waste. The recyclability of the materials should, therefore, also be considered when designing sustainable rubbers. Unfortunately, the chemical reactions typically used to crosslink elastomers are irreversible and, thus, prohibit reuse of rubber scrap and waste as raw materials. Thermoplastics can be recycled *via* melt processing, but the three-dimensional (3D) network of crosslinked rubbers prevents melt (re)processing. This problem is particularly evident for rubber tyres of which, due to the inability of recycling them, millions are discarded annually [5]. The bulk of these are dumped in landfills, placing a burden on the environment as well as posing potential health hazards because these become a breeding ground for disease-carrying mosquitoes.

Current trends towards sustainability and the development of products in a cradle-to-cradle fashion (a biomimetic approach to the design of products and systems that models human industry on nature's processes viewing materials as nutrients circulating in healthy, safe metabolisms) make the recyclability of crosslinked elastomers increasingly relevant [6]. In the last decades, considerable efforts have been devoted to the devulcanisation of various crosslinked rubbers [7–12]. For some iso-based rubbers, such as NR and butyl rubber (BR), reclaiming processes have been commercially practiced for decades. Also, reclaiming sulfur- vulcanised rubbers using devulcanising agents and high-shear/ temperature processes is now a common technology [13–15]. It appears to be more difficult to apply this technology to hydrocarbon elastomers with saturated main chains [16–18], probably because of the higher stability of sulfur crosslinks in an environment with low unsaturation. Nevertheless, some workable reclaiming processes have been developed for these materials [7, 17]. These processes combine thermal and mechanical treatment for selective cleavage of the sulfur crosslinks. Unfortunately, they also cause a considerable amount of scission of the main polymer chain, which is detrimental for the performance of the recyclate [7]. Hence, the amount of devulcanised material that can be reused in new products is limited to ≈25% [17].

A technical solution to this problem is found in the commercially available thermoplastic elastomers (TPE). TPE comprise block copolymers of 'hard' and 'soft' segments, and thermoplastic vulcanisates (TPV) that consist of blends of 'hard' and 'soft' polymers [19–21]. TPE and TPV show advantages typical of rubbery materials and plastic materials, such as easy processing and manufacture.

Unfortunately, these materials have a relatively low stability at high temperatures that limits their applicability. Newly developed alternative solutions are discussed in Section 1.2.

Combining green processing with the use of bio-based materials and a recyclable end-product would ultimately lead to a sustainable rubber product. However, recycling is not the most favourable way to deal with used rubber products, and it would be more desirable to prevent the rubber from becoming waste in the first place. A 'smart' solution for rubbers can, therefore, be found in 'self-healing polymers' [22], a category of smart rubbers that can repair or heal any inflicted damage, as discussed in Section 1.3.

1.1 Bio-Based rubbers

The use of bio-based rubbers would result in a lower dependence on fossil fuels. Currently however, ever more NR is being replaced by synthetic rubbers. NR has been used since before the 19th century, mostly by the indigenous tribes in the Brazilian rain forests [23]. In the late-19th century, the discovery of its existence and applications by the industrialising world gathered momentum, and the production and use of NR started picking up. Shortly before and during World War I, the production increased strongly due to the need for rubbers in many novel applications [24]. Rubbers were required for many products that were suddenly mass-produced as well as for the machinery used to make them. Due to fluctuations in reliable production of NR, the supply was unable to keep up with the rapidly growing demand. Therefore, an alternative was sought for and found in 1909 by German scientists: oil-based synthetic rubbers [25]. The development of the first synthetic rubbers led to a boost in further development of these materials [26]. Since then and throughout the 20th century, an expanding range of synthetic rubbers has been invented.

The main shortcoming of this broad range of synthetic rubbers is their resistive properties. NR is, in general, susceptible to degradation upon exposure to sunlight, oxygen, heat and ozone due to a relatively high degree of unsaturated carbon–carbon bonds in the polymer backbone of cis-polyisoprene. Bio-based alternatives for more saturated, synthetic rubbers should, therefore, be developed to allow for an increased use of bio-based rubbers.

An obvious solution for the relatively low resistive properties of NR with respect to synthetic elastomers with a higher degree of saturation is the hydrogenation of NR [27, 28]. Saturating the double bonds through hydrogenation improves the stability of NR while retaining its bio-based character. Another example of reducing the amount of unsaturated carbon–carbon bonds in the rubber backbone is the copolymerisation of methyl methacrylate onto the NR latex [29]. The thermal, mechanical, ozone-ageing, and solvent-resistance properties of the resulting modified NR have been found to be superior to those of unmodified NR [29]. The decomposition temperature of the grafted/hydrogenated NR was found to be higher than that of the unmodified rubber by $\approx 30\,°C$, and the mechanical strength was also improved.

Another approach that could lead to an increase in the bio-based content of used rubber products is replacing the monomer building blocks with bio-based analogues. A good example is the recently developed, partly bio-based ethylene propylene diene rubber (EPDM) [30]. EPDM rubbers usually consist of ethylene and propylene produced *via* cracking of natural oil. To produce bio-based EPDM, the ethylene was ultimately derived from sugar cane. The sugar within sugar cane was first converted to ethanol, dehydrated to ethylene and then used in a solution-based Ziegler–Natta polymerisation to yield EPDM rubbers with a bio-based ethylene content of 50–70 wt%. The resulting gum rubbers were found to have exactly the same material properties as the synthetic analogue [31].

Typically, rubber products do not consist only of elastomers, but also contain significant amounts of (reinforcing) fillers [typically carbon black(s) (CB), silica or inert white fillers], plasticisers, crosslinking agents and other additives. EPDM products, for example, may contain ≤400 parts per hundred (phr) rubber of compounding ingredients [32]. The majority of such rubber compounds, in general, consist of additives. Hence, the sustainability of these additives should also be considered, including thinking about alternatives for traditional plasticiser oils and (reinforcing) fillers, which are, in general, not sustainable. CB, for example, is typically produced through the incomplete combustion of hydrocarbons with natural gas. Silica is produced *via* precipitation from a silicate salt solution, and inert white fillers are extracted from mines and milled to fine powders. Typically, traditional plasticisers are refinery fractions of crude oil. Using sustainable alternatives for traditional plasticiser oils and (reinforcing) fillers are required for the development of rubber compounds with the highest sustainable content.

Recycled CB produced by the pyrolysis of waste rubber tyres appears to be a suitable, sustainable alternative for certain traditional medium-reinforcing CB fillers [33]. These pyrolysis CB are a more sustainable ingredient because rubber waste is considered to be a major environmental issue. Also, pyrolysis seems to be one of the preferred recycling technologies that reduce the CO_2 production by 5 tonnes per tonne of rubber compound [34]. Silica could be replaced with rice husk ash, which is the silica recovered *via* burning off the organic fraction of rice husk obtained during rice cleaning [35]. Natural fibres (jute, palm, sisal, hemp) or natural flours and powders (wood, cork, soy) could be suitable sustainable alternatives for inert, mineral fillers [36–39] because they do not contribute to the compound performance, they only dilute it. Unfortunately, it appears that these bio-based fillers result in poor physical performance [32, 34], probably because of their lower surface area combined with a lack of reactivity. Finally, unsaturated natural oils and fats, such as palm, rice bran, ground nut, soybean, mustard and sunflower oils, have been tested as plasticisers in apolar rubber products [40–42]. Unfortunately, their high polarity and/or high levels of unsaturation typically results in poor compatibility (mixing issues and oil bleeding) and competition for sulfur vulcanisation (resulting in a reduced crosslink density and corresponding inferior vulcanisate properties). Squalane appears to be a better bio-based alternative for mineral oil plasticisers for apolar rubbers because it is also apolar and fully saturated [32].

1.2 Recyclable and self-healing rubbers

1.2.1 Recycling rubber

Non-crosslinked elastomers are sticky, chewing gum-like materials that do not possess the elasticity, resilience, strength and/or dimensional stability required for rubber applications. Crosslinking is an absolute requirement for the prompt and complete recovery from deformation (i.e., for the elastic recovery of rubber materials) [10]. Moreover, crosslinking enhances other properties, such as strength, temperature stability, and chemical and stress-cracking resistance [13]. Unfortunately, the chemical reactions typically used to crosslink elastomers are irreversible and, thus, prohibit reuse of rubber scrap and waste as a raw material. Whereas thermoplastics can simply be recycled *via* melt processing, the 3D network of crosslinked rubbers prevents melt (re)processing. This problem is particularly evident for rubber tyres of which, due to the inability of recycling them, millions are discarded annually [5]. The majority of these are dumped in landfills, placing a burden on the environment as well as posing potential health hazards as these become a breeding ground for disease carrying mosquitoes. Current trends towards sustainability and cradle-to-cradle products make the recyclability of crosslinked elastomers increasingly relevant.

In the last decades a lot of effort has, therefore, been made to devulcanise rubber [14–18]. Reclaiming of sulfur-vulcanised NR, using devulcanising agents and high shear/temperature processes, is now a common technology [13, 43, 44]. It appears to be more difficult to apply this technology to hydrocarbon elastomers with a saturated backbone, such as EPDM. This is probably related to the slightly higher stability of sulfur crosslinks in an environment with low unsaturation. Established processes for such saturated elastomers make use of devulcanising agents and high-shear/ temperature processes in an extruder and enabling the reprocessing used rubber to such an extent certain that, when it is mixed with 50% virgin material, a product with the similar material properties is obtained [16, 18].

1.2.2 Reversible crosslinking

A complete cradle-to-cradle alternative for the recycling of rubbers *via* de-crosslinking would require a material that combines the material properties of a permanently crosslinked rubber with the recyclability of a non-crosslinked thermoplastic. A technical solution is found in the commercially available TPE, consisting of block copolymers of hard and soft segments, and TPV, consisting of blends of hard and soft polymers [20, 21]. Another approach to achieve recyclability is *via* rubber networks with reversible crosslinks that respond to an external stimulus, such as temperature. The formation of crosslinks at relatively low temperatures is beneficial for good elastic and mechanical performance, whereas the cleavage of crosslinks at high temperatures (similar to the processing temperature of the original, non-crosslinked rubber compound) allows for recycling of the rubber product. Several approaches have been developed for the thermo- reversible crosslinking of existing rubbers (Figure 1.3) [5, 45–49]. Most of these approaches use thermo-reversible chemical reactions to introduce crosslinks that can open at temperatures above the application temperature and below the degradation temperature of the polymer.

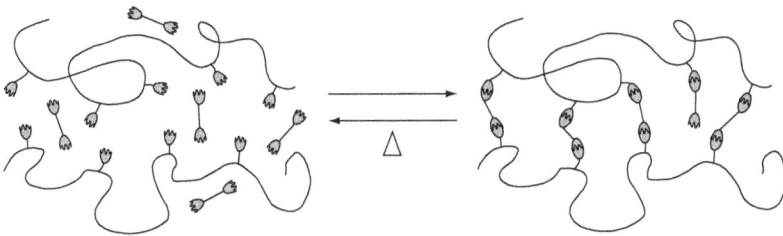

Figure 1.3: Thermo-reversible crosslinking of a polymer using bifunctional crosslinkers as 'bridges' between functional groups on the polymer backbone (schematic).

Thermally-reversible crosslinks broadly fit into 'covalent' and 'non- covalent' interactions. Non-covalent or 'physical' interactions are, in general, applied for thermo-reversible crosslinking and are relatively weak interactions [48, 50]. Examples of such non-covalent reversible crosslinks are hydrogen bonds [51–53], electrostatic interactions [53–55] and van der Waals interactions (e.g., styrene-butadiene- styrene rubber). Unfortunately, such thermo-reversibly 'crosslinked' rubbers cannot indefinitely hold stress without creep and also have limited applicability at elevated temperatures [56]. Reversible chemical crosslinks are based on the covalent bonds that can be broken by external influences. These stronger, reversible covalent crosslinks can be achieved by, for instance, thermo-activated di/ polysulfide rearrangements [57], reversible sulfur bridges of special cases of thermally-reversible ester bonds, and amide bonds. These reversible crosslinks will, in general, result in enhanced mechanical performance [58–60]. Although several other covalent, thermo- reversible crosslinking reactions are known, the Diels–Alder (DA) reaction is undoubtedly the most frequently used. Recently, several approaches have been developed that employ the retro Diels–Alder (rDA) reaction for the thermo-reversible crosslinking of different polymers [47, 61–63] and elastomers [49, 64, 65]. These DA crosslinked rubbers show material properties similar to those of conventionally crosslinked rubbers and can be reprocessed with high retention of properties [49, 66]. Furthermore, these elastomers remain reprocessable after compounding them with CB fillers and oil plasticisers [67].

1.2.3 Self-healing elastomers

In biological systems, the original function can be restored at macroscopic (healing blood vessels or broken bones) and molecular level (e.g., repair of deoxyribonucleic acid) upon infliction of minor damage (if non-fatal). Conversely, polymers often have a limited lifespan due to unavoidable degradation and unexpected damage from constant stress and strain. To overcome these issues, self-healing polymers defined as 'materials where damage automates a healing response' have been developed. Smart rubbers are designed so that they heal (recover/repair) damage automatically and autonomously at ambient conditions, that is, without any external intervention [68, 69] through a succeeding autonomous process [22]. Chemical (reversible and polymeric) and non-chemical (irreversible and microvascular) systems can be employed for the development of self- healing elastomers. The main benefit of application of these materials is that they may not have to be replaced. As such, the development of self-healing rubbers is relevant in diverse areas. Self-healing rubbers would be particularly useful in applications where accessibility is difficult and/or high-level reliability is required. Different strategies have been developed for the design of self-healing materials, but the underlying principle is always the same (Figure 1.4). After damage is inflicted by, for instance, a mechanical load which creates a crack (Figure 1.4A), a trigger generates a 'mobile phase' (Figure 1.4B). The latter is then transported to the crack site, where it mends the two crack planes by physical interactions and/or chemical bonds (Figure 1.4C). Once healing is completed, the mobile phase is immobilised again (Figure 1.4D) and the material properties are restored completely.

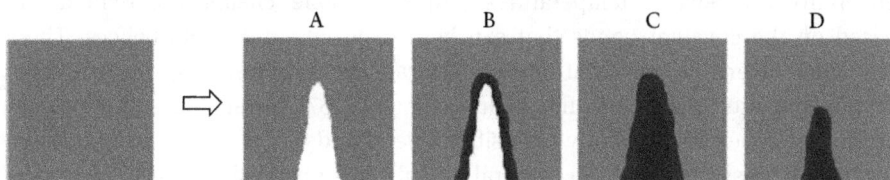

Figure 1.4: The general self-healing mechanism in non-biological materials. (A) A crack is formed by a mechanical load and (B) a 'mobile phase' generated. (C) The mobile phase closes the crack and (D) is immobilised. Adapted and redrawn from M.D. Hager and U.S. Schubert in *Self-Healing at the Nanoscale: Mechanisms and Key Concepts of Natural and Artificial Systems*, Eds., M. Amendola and M. Meneghetti, CRC Press, Boca Raton, FL, USA, 2012, p.291 [70].

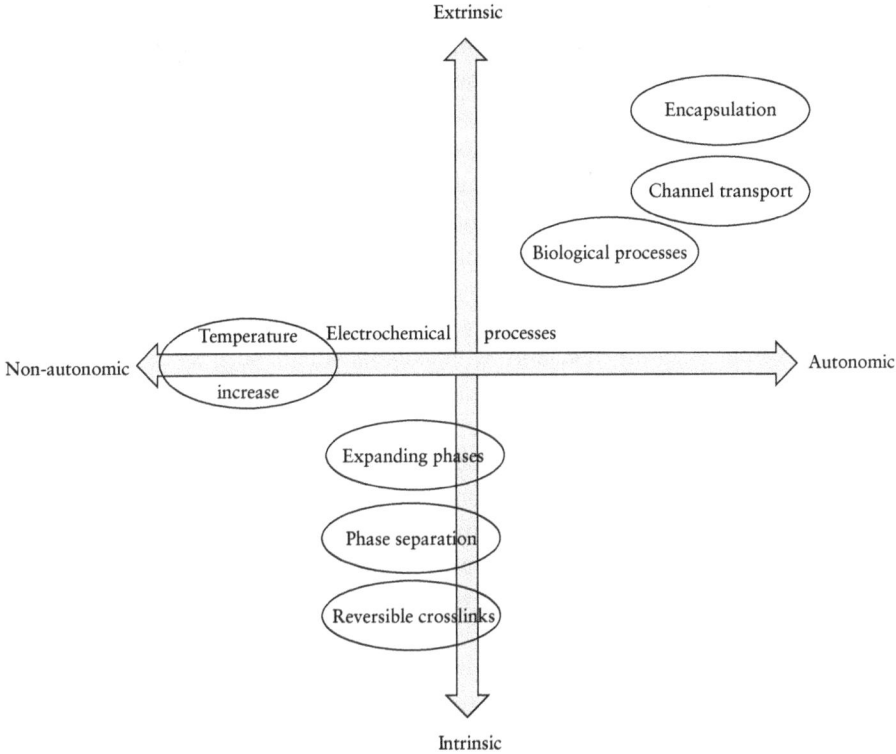

Figure 1.5: Classification of healing mechanisms with respect to extrinsic and intrinsic as well as autonomic and non-autonomic self-healing. Adapted and redrawn from M.D. Hager and U.S. Schubert in *Self-Healing at the Nanoscale: Mechanisms and Key Concepts of Natural and Artificial Systems*, Eds., M. Amendola and M. Meneghetti, CRC Press, Boca Raton, FL, USA, 2012, p.291 [70].

The way in which this self-healing takes place can be distinguished into three categories (Figure 1.5). First, a distinction can be made based on whether a trigger to start self-healing is required. Such a trigger can be internal (thus, only induced by the damage itself) or external (e.g., by heat or light). Autonomic self-healing materials can heal without any intervention (thus the damage is the stimulus), whereas non-autonomic self-healing materials require an external trigger to initiate the self-healing process. The second distinction is the absence or presence of a healing agent. Intrinsic healing materials heal themselves without the addition of a healing agent. Extrinsic healing materials have pre-embedded healing agents. Intrinsic healing can be realised by chemical and/or physical interactions, whereas extrinsic healing can be achieved by the addition of tubes or microcapsules loaded with healing agent throughout the material. Extrinsic methods are also autonomic whereas most intrinsic methods are non-automatic. Furthermore, extrinsic methods are usually suitable to self-heal the material once, whereas intrinsic methods could, in principle, self-heal the material indefinitely. Finally, a distinction can be made between full-scale restoration to the initial state and functionality restoration, which restores only the principal function.

Smart materials are often prepared by combining state-of-the- art engineering techniques with efficient chemical reactions such as 'click chemistry'. Other systems make use of microcapsules or microvascular systems and nano-reservoirs which contain a crosslinker, catalyst or monomer and crack open upon imposed damage to initiate a polymerisation reaction that blocks or heals the crack. These irreversible systems can heal only once, so there is increasing interest in reversible systems, which can break and heal repeatedly if triggered by external stimuli [71, 72]. In such systems, the interaction of chemical functionalities with external stimuli such as thermal, electrical, pressure, mechanical and electrochemical radiation should repair the damage at the microscopic level [72, 73]. Such reversible systems can be distinguished on the basis of the nature of the interaction. Some of these intrinsic healing systems make use of reversible physical interactions (non-covalent), such as hydrogen bonding, π–π stacking, electrostatic interactions, or van der Waals interactions. Other systems are based on reversible covalent interactions or crosslinks, making use of DA and rDA processes, photo-reversible crosslinking and reversible disulfide crosslinking.

1.2.4 Autonomous self-healing by damage-triggered smart containers

The introduction of microcapsules containing liquid monomers in the polymer matrix can be used to facilitate the autonomous repair of cracks [74]. In such cases, the polymerisation catalyst particles are usually also dispersed throughout the polymer matrix. The idea is that if a crack in the polymer occurs, the local capsules rupture, causing the monomer to flow into and fill the crack. Subsequently, the released monomers are polymerised by the catalyst (Figure 1.6).

Four basic criteria must be met for the successful implementation of this method: storage, release, transport, and rebonding. The healing agent and catalyst should remain stable for long periods of time until a crack occurs. The capsules must be strong enough to survive the processing of the material, but weak enough to open if a crack occurs. To ensure appropriate transport from the monomer to the crack, the liquid must have a lower surface energy than the fracture, but must not be too volatile to prevent evaporation or diffusion from the crack plane. If the crack is filled with monomer, the catalyst should be able to start rapid polymerisation at room temperature (RT) [75]. Some good examples of the successful implementation of this method of self-healing make use of the ring-opening metathesis polymerisation reaction with, for example, di-cyclopentadiene as an encapsulated monomer [76]. Polycondensation [77] and epoxy-based [78] healing reactions are also frequently employed by encapsulating both reagents separately and having them recombine in the cracks. The healing efficiency of these systems has increased from 75% in 48 h [76] to 90% in

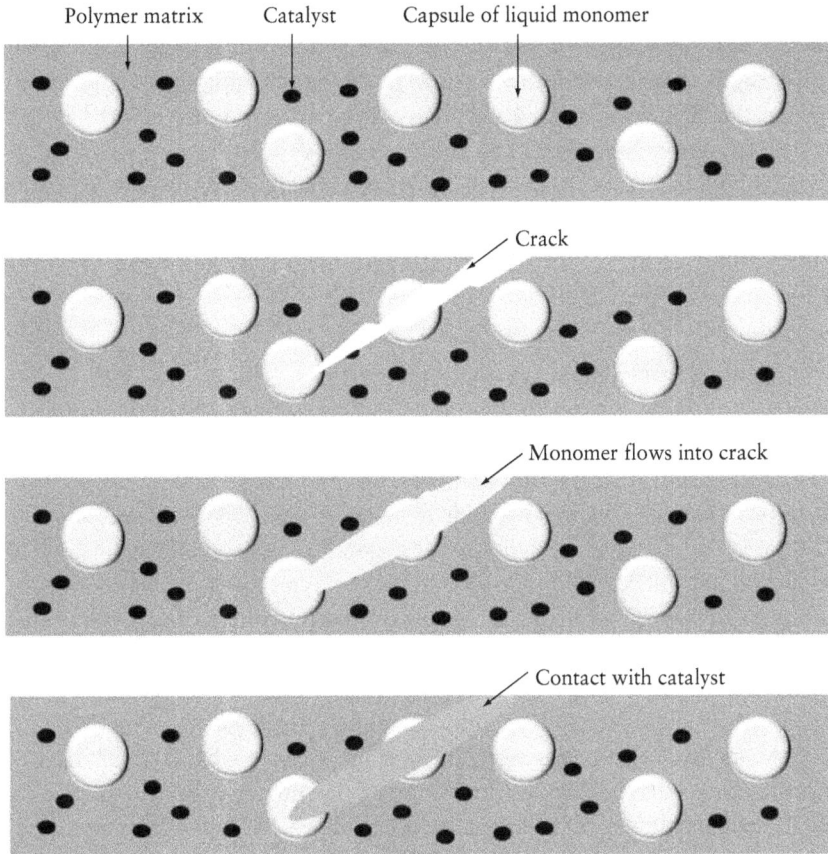

Figure 1.6: Mechanism for autonomous crack healing with microcapsules containing monomers that polymerise if brought into contact with the appropriate catalyst. Adapted and redrawn from M.R. Kessler, *Periodical Proceedings of the Institution of Mechanical Engineers, Part G: Journal of Aerospace Engineering*, 2007, **221**, G4, 479 [75].

10 h [79] over the last years. This even led to an ultra-fast self-healing system that yields high healing efficiencies (>70%) within seconds [80]. Shortcomings of these systems are that only small cracks can be repaired because of the limited amount of monomer that can be stored in the microcapsules. Hence, the material can only be healed once at a specific spot, and that the healing efficiency is reduced over time, resulting in low stability [74].

1.2.5 Intrinsic self-healing via reversible crosslinking

Reversible crosslinking represents one of the most promising and studied fields in polymer chemistry [48]. However, most systems making use of reversible crosslinks are non-automatic and require an external trigger to heal the material or repair any imposed damage. Although external stimuli such as electricity, magnetism, shear stress, pH or light can be used for the control of the crosslinking process, temperature remains the most investigated and used external trigger for such processes [48]. A thermally-reversible crosslinked polymer can be processed as a thermoplastic material at high temperature, whereas the material will behave like a crosslinked network at lower temperatures. Even though not all the crosslinks are broken during the process, the reduced number of crosslinks provides enough thermoplasticity to the polymers for reworking or self-healing.

1.2.5.1 Self-healing through non-covalent interactions

One of the most exploited ways to accomplish self-healing is by using reversible, non-covalent or physical interactions in the form of hydrogen bonds (the attractive interaction between a hydrogen atom with an electronegative one, such as nitrogen, oxygen or fluorine). Hydrogen bonds have an energy of 5–30 kJ/mol and are stronger than van der Waals interactions, but generally weaker than electrostatic or covalent bonds and will, therefore, break more easily upon the application of stress. Hence, if damage is inflicted onto a material with both hydrogen bonds and covalent bonds, predominantly the hydrogen bonds will break whereas the covalent bonds will remain intact. Hydrogen bonds are reversible interactions, so they are restored afterwards automatically, resulting in the self-healing of the material.

The first time hydrogen bonds were used for thermo-reversible crosslinking was in a polysiloxane with carboxylic pendent groups [81]. Although the crosslinks were stable at RT, some side reactions started to occur above 70 °C. Furthermore, water appeared to facilitate the rupture of the network. A more recently developed thermo-reversible isoprene rubber with triazole rings clearly yielded a crosslinked rubber with a second endothermic transition at ≈185 °C [82]. The tensile strength of this isoprene rubber with 3.8 mol% of pendant triazine groups was found to be much higher than that of uncured isoprene rubber containing 30 phr of CB and comparable with that of sulfur-vulcanised rubber [82]. BR was also thermo- reversibly crosslinked using hydrogen bonds by introducing a methyl phenyl sulfone carbamate, and the glass transition temperature (T_g) increased to 100 °C upon modification with 20 mol% [52]. This thermo-reversibly crosslinked BR was also filled with silica particles, modified with phenyl (no hydrogen-bonding interaction) and OH groups (hydrogen-bonding interaction) [83]. It was found that the hydrogen bonding interaction between the filler and rubber was more sensitive to temperature than the hydrogen bonding interaction between the filler and filler. As a result, mixing with 40 wt% of the silica particles modified with OH resulted in a relatively higher reinforcement and temperature stability with respect to the silica particles modified with phenyl groups [83]. Finally, a maleated ethylene propylene rubber copolymer (EPM) was modified with an excess of an amino compound, which appeared to form a supramolecular structure [84]. The mechanical properties of the resulting thermo-reversibly crosslinked EPM network was dependent upon the aliphatic chain in the amine compound. It was found that longer apolar tails disturbed aggregate formation, leading to worse properties. Simultaneously, longer tails may also organise themselves in a crystalline-like order, which will improve the properties [84]. An important concern for such systems is imide formation of the amide acids and amide salts upon compression moulding at temperatures above 120 °C, resulting in disappearance of the aggregates and poor mechanical properties [48].

The retention of material properties is usually taken as a measure for the efficiency of self-healing. Several characteristic material properties, such as rheological properties (to determine the shear storage modulus and shear loss modulus) [56, 85–91] and tensile properties (to determine the Young's modulus, toughness, ultimate tensile strength and elongation at break) [56, 87–89, 91, 92] are typically measured to determine the performance of elastomers before and after self-healing. Tensile test measurements on thermo-reversibly crosslinked rubbers with hydrogen bonds have shown that the stress and strain recovery (retention of material properties) increase with the time the material has had to self-heal. The conditions at which this self-healing takes place also have important roles. The self-healing behaviour of modified block copolymers with poly(n-butyl acrylate) as the soft phase and polystyrene (PS) as the hard phase is a good example [89]. If the covalent bonds in the centre of the poly(n-butyl acrylate) are replaced with a quadruple hydrogen-bonding junction, the material becomes self-healing. 2-Ureido-4-pyrimidinone is a suitable quadruple hydrogen bonding motif because it has high thermodynamic stability and rapid kinetic reversibility [93]. The kinetic reversibility in deuterated chloroform at 26.85 °C has been reported to be 8 s^{-1} [89] whereas the kinetic reversibility in deuterated water and toluene are 13 and 0.6 s^{-1}, respectively [93]. The healing efficiency of self-healing polymers with hydrogen bonds can be increased by the introduction of aromatic disulfides as crosslinkers [92]. This approach is related to the dynamic nature of the aromatic disulfides at RT. Another way to increase self-healing kinetics is by using complementary hydrogen bonding groups, such as amidoethyl imidazolidone, di (amidoethyl) urea and diamido tetraethyl triurea [56]. The number of available donors and acceptors of hydrogen bonds at the moment of mending, in general, decreases with time because the mending of damage over time is related to the slowness of the re-association of hydrogen bonds. Therefore, a self-healing material has a slow re-association (where it leaves hydrogen-bond donors and acceptors available for mending after the waiting time) or a self-healing material has a fast re-association with a reduced ability of mending after the waiting time. This would result in a trade-off between the self-healing speed and ability of mending after the waiting time. In general, it has been found that the self-healing speed is dependent on the available hydrogen-bond acceptors and donors and on the mobility of the groups [87]. If the density of available hydrogen-bond acceptors and donors at the cut interface is increased, the self-healing speed is expected to increase as well. Thus, if the cut parts are not joined immediately, groups within each broken part also recombine faster and leave fewer groups available for bonding between the two cut parts upon mending over time, resulting in less recovery (more decay). Finally, the effect of the polymer architecture on the properties and self-healing ability of elastomers has been demonstrated using an autonomic self-healing TPE with a soft phase (polyacrylate amide brushes) and a hard phase (PS backbone) [88]. For this elastomer, increasing the weight fraction of styrene from 3 to 7% resulted in an increase in the Young's modulus by a factor of three without compromising the material's self-healing ability.

The interaction between ionomers is, in general, stronger than that for hydrogen bonds and also frequently used in the design of self-healing rubbers. Ionomers are a class of polymers that contain small numbers (≤15 mol%) of ionic groups neutralised with a metal ion, pendant from or incorporated into a hydrocarbon backbone [94]. These materials are, in general, applied in shoes or food packaging because of their high transparency, toughness, flexibility, adhesion, and oil resistance [5]. The position of the ionic groups, the type of groups and counterions attached to the backbone are the main variables that can be used to 'tune' the material properties of ionomers [95]. The thermal reversibility of ionomers may be hindered because they also display a certain degree of shape-memory due to the formation of ionic domains [96–99]. Nevertheless, a wide variety of self-healing elastomers has been developed on the basis of ionomeric interactions [54, 55, 81, 100–103]. The melt processability of ionomers is a result of the increased mobility of ionic pairs inside the aggregates at elevated temperatures.

Electrostatic interactions can also be used for the reinforcement of elastomers that are already reprocessable, such as TPE. Metal- neutralised sulfonated EPDM rubbers, for example, display mechanical properties comparable with standard vulcanised EPDM after the incorporation of electrostatic crosslinks [104–106]. To avoid degradation, however, sulfonated ionomers must be fully neutralised. This disadvantage explains why carboxylated ionomers are usually preferred [5]. Another example is the sulfonation of some of the aromatic rings of thermoplastic ionomers based on styrene-grafted NR [107]. The addition of ionic groups improves the tensile strength by a factor of 10 and the resulting material is shown to be completely reprocessable for ≥3 cycles without loss of mechanical properties. The balance between the molecular weight of the polymer and the grafting density of ionic groups also allows the control of TPE processing properties [5]. Self-healing properties have also been obtained by ionic modification of bromobutyl rubber [108]. Finally, maleated EPM rubbers have been thermo-reversibly crosslinked using electrostatic interactions to display material properties comparable with those of peroxide-cured elastomers [54, 55].

Even though non-covalent interactions, such as hydrogen bonds or electrostatic interactions, have the advantage of easy reversibility thanks to their low bond energy, such non-covalently crosslinked materials barely succeed in equalling co-valently bonded elastomers in terms of thermomechanical properties [5]. To reach high tensile strength, elasticity or low compression set, covalent crosslinks are undoubtedly more favourable. However, covalent chemistry is inconsistent with recycling, unless the covalent crosslinks are reversible bonds.

1.2.5.2 Thermo-reversible chemical interactions

When considering chemical interactions for the thermo-reversible crosslinking of elastomers, the DA and rDA reactions are particularly interesting because they are addition processes. Hence, all atoms of the starting components will be found in the product, such that nothing is consumed and no by-products are formed during the process [5]. A classical approach for the thermo-reversible crosslinking of elastomers using DA chemistry is combining a functionalised polymer (containing a diene or dienophile side groups) with a small bifunctional molecule [*bis*(maleimide) or difuran] as a crosslinker. Fully recyclable, low-T_g elastomers have been prepared on the basis of siloxane rubber [61], BR [64, 109], NR [61] and maleated EPM [49] using this approach. The resulting materials can be cut and compression-moulded back to the original shape and recover most of their mechanical properties. The large number of diene and dienophile candidates that are available with specific temperatures for the crosslinking and de-crosslinking reactions makes this system very versatile [48]. The reaction between furan and maleimide, for example, is known to be an equilibrium in favour of adduct formation under 60 °C and of its dissociation above 100 °C [48].

Other reactions can also be used to create covalent networks that can adapt in ambient conditions, allowing the relaxation of stress and enabling the self-healing of the material. Reversibly crosslinked elastomers using sulfur chemistry have been reported widely [110–113]. An example is the metathesis reaction of aromatic disulfides. Poly (urea-urethane) elastomeric networks have been demonstrated to mend completely after being cut in two parts by a razor blade *via* thermal annealing [92, 114]. Another example of self-healing rubbers is found in epoxidised NR crosslinked with carboxylic diacids through an esterification reaction with epoxy functions [115, 116]. The resulting β-hydroxy esters compete with the classical sulfur or peroxide vulcanisation crosslinking chemistry usually employed for the crosslinking of elastomers [117, 118]. A major benefit of this system is that the diacid used (e.g., dodecanedioic acid) can be obtained from renewable resources [119], making this system fully bio-based.

A final example of self-healing elastomers is found in vitrimers [120]. The latter are a new class of polymers that are fully malleable and recyclable, but stay insoluble in any solvent even under heating. This behaviour is a result of covalent exchangeable bonds making use of reactions, such as transesterification [120–122], olefin metathesis [123], vinylogous *trans*-amination [124] or transalkylation, that are used as crosslinks. Due to exchange chemistry, the number of crosslinking points and average functionality of the network remain constant, making vitrimers fully malleable and mendable (but still insoluble) materials [120, 122]. The topology of the network is quenched by cooling, which fixes the shape of the material by slowing down exchange reactions, resulting in a practical 'freezing' of the transesterification reactions [56, 120, 121]. The resulting materials can be reshaped and reprocessed repeatedly and combine the good mechanical properties of covalently crosslinked elastomers with full recyclability. Vitrimer chemistry has already been applied to BR [123] and could also be applicable to all olefin-containing rubbers, such as NR and styrene-butadiene rubber [5].

1.3 Concluding remarks

Sustainability is a crucial aspect in the design of smart rubbers. This is especially true when considering their carbon footprint and recyclability. The carbon footprint of rubbers is reduced greatly if using NR or synthetic rubbers derived from bio-based materials such as EPM from bio-ethanol. Most rubber products do not, in general, consist of the elastomer, so bio-based and greener alternatives for the conventionally used CB fillers and oil lubricants are suggested.

Unfortunately, even if rubber products are completely bio-based, the used products remain un-reprocessable and will eventually end up as rubber waste, causing major environmental issues. Several more sustainable options are proposed for the end-of-life management of rubber products. TPE and TPV are practical solutions applied in several rubber applications, but cannot match the rubber properties for some more demanding applications. Reversible covalent crosslinking of rubbers could be a solution for the development of similar materials with enhanced high-temperature performance and lower creep. An even more desired solution is found in the design of smart rubbers with self-healing properties. The chemistry used for intrinsic self-healing polymers is also largely applicable for thermo-reversible crosslinking. Although it is challenging to develop materials that display self-healing behaviour at service conditions while retaining mechanical and physical properties at these same conditions, some newly developed chemistry has been suggested for the dynamic crosslinking of elastomers. A promising example is found in vitrimer chemistry applied to industrial rubbers that combine recyclability with full solvent resistance. A disadvantage of these approaches is the necessity of designing and synthesising new polymers. This approach may hamper the substitution of currently employed materials. Modification of existing polymers would give the most facile implementation because the resulting materials are similar to the currently employed materials.

References

1.   F. Picchioni, *Polymer International*, 2014, **63**, 8, 1394.
2.   T. Amnuaysin, P. Buahom and S. Areerat, *Journal of Cellular Plastics*, 2016, **52**, 6, 585.
3.   S.K. De and J.L. White in *Rubber Technologist's Handbook*, Smithers Rapra, Shawbury, UK, 2001.
4.   W. Hofmann in *Rubber Technology Handbook*, Hanser Publishers, Münich, Germany, 1989.

5. L. Imbernon and S. Norvez, *European Polymer Journal*, 2016, **82**, 347.
6. M. Braungart and W. McDonough in *Cradle to Cradle: Remaking the Way We Make Things*, North Point Press, New York, NY, USA, 2002.
7. K.A.J. Dijkhuis, I. Babu, J.S. Lopulissa, J.W.M. Noordermeer and W.K. Dierkes, *Rubber Chemistry and Technology*, 2008, **81**, 2, 190.
8. M.A.L. Verbruggen in *Devulcanization of EPDM Rubber*, University of Twente, Enschede, The Netherlands, 2007. [PhD Thesis]
9. I. Mangili, E. Collina, M. Anzano, D. Pitea and M. Lasagni, *Polymer Degradation and Stability*, 2014, **102**, 15.
10. J. Shi, K. Jiang, H. Zou, L. Ding, X. Zhang, X. Li, L. Zhang and D. Ren, *Journal of Applied Polymer Science*, 2014, **131**, 11, 40298.
11. B. Adhikari, D. De and S. Maiti, *Progress in Polymer Science*, 2000, **25**, 7, 909.
12. W.C. Warner, *Rubber Chemistry and Technology*, 1994, **67**, 3, 559.
13. G.K. Jana and C.K. Das, *Polymer-Plastics Technology and Engineering*, 2005, **44**, 2, 1399.
14. G.K. Jana, R.N. Mahaling, T. Rath, A. Kozlowska, M. Kozlowski and C.K. Das, *Polimery*, 2007, **52**, 2, 131.
15. M. Myhre, S. Saiwari, W. Dierkes and J. Noordermeer, *Rubber Chemistry and Technology*, 2012, **85**, 3, 408.
16. P. Sutanto in *Development of a Continuous Process for EPDM Devulcanization in an Extruder*, University of Groningen, Groningen, The Netherlands, 2006. [PhD Thesis]
17. P. Sutanto, F. Picchioni, L.P.B.M. Janssen, K.A.J. Dijkhuis, W.K. Dierkes and J.W.M. Noordermeer, *Journal of Applied Polymer Science*, 2006, **102**, 6, 5948.
18. P. Sutanto, E. Picchioni, L.P.B.M. Janssen, K.A.J. Dijkhuis, W.K. Dierkes and J.W.M. Noordermeer, *International Polymer Processing*, 2006, **21**, 2, 211.
19. M. Maiti, A. Bandyopadhyay and A.K. Bhowmick, *Journal of Applied Polymer Science*, 2006, **99**, 4, 1645.
20. A.Y. Coran and R.P. Patel in *Thermoplastic Elastomers*, 2nd Edition, Eds., G. Holden, N.R. Legge and H.E. Schroeder, Hanser Publications, Cincinnatti, OH, USA, 1996.
21. J. Karger-Kocsis in *Polymer Blends and Alloys*, Eds., G.O. Shonaike and P. Simon, CRC Press, Boca Raton, FL, USA, 1999.
22. S. van der Zwaag in *Self-Healing Materials*, Springer, New York, NY, USA, 2007.
23. H.H. Greve in *Ullmann's Encyclopedia of Industrial Chemistry*, Ed., B. Elvers, John Wiley and Sons, Inc., Hoboken, NJ, USA, 2000.
24. W. Dean in *Brazil and the Struggle for Rubber: A Study in Environmental History*, Cambridge University Press, Cambridge, UK, 1987.
25. M. Michalovic in *Destination Germany: A Poor Substitute*, Polymer Science Learning Center, 2000.
26. D. Threadingham, W. Obrecht, J.P. Lambert, M. Happ, C. Oppenheimer-Stix, J. Dunn, R. Krüger, H. Brandt, W. Nentwig, N. Rooney, R.T. LaFlair, U.U. Wolf, J. Duffy, J.E. Puskas, G.J. Wilson, H. Meisenheimer, R. Steiger, A. Marbach, K.M. Diedrich, J. Ackermann, D. Wrobel, U. Hoffmann, H.D. Thomas, R. Engehausen, S.D. Pask, H. Buding, A. Ostrowicki and B. Stollfuss in *Ullmann's Encyclopedia of Industrial Chemistry*, Ed., B. Elvers, John Wiley and Sons, Inc., Hoboken, NJ, USA, 2000.
27. C. Kookarinrat and P. Paoprasert, *Iranian Polymer Journal*, 2015, **24**, 2, 123.
28. N.T. Ha, K. Shiobara, Y. Yamamoto, L. Fukuhara, P.T. Nghia and S. Kawahara, *Polymers for Advanced Technologies*, 2015, **26**, 12, 1504.
29. N. Jamaluddin, I. Abdullah and S.F.M. Yusoff, *AIP Conference Proceedings*, 2015, **1678**, 1, 050036.
30. U. Erbstoesser, *Kautschuk Gummi und Kunststoffe*, 2014, **67**, 3, 12.

31. M.A. Grima, P. Hough, D. Taylor and M. van Urk, *European Rubber Journal*, 2013, 28.
32. M. van Duin and P. Hough in *Proceedings of the DKT/IRC2015*, Nuremburg, Germany, 2015.
33. R. Leunissen, *Tyre and Rubber Recycling*, 2016, 19.
34. M. van Duin and P. Hough, *Kautschuk, Gummi, Kunststoffe*, 2017. [In Progress]
35. S. Ragunathan, H. Ismail and K. Hussin, *Journal of Thermoplastic Composite Materials*, 2014, **27**, 12, 1651.
36. K.A. Job, *Rubber World*, 2014, **March**, 32.
37. E.M. Fernandes, V.M. Correlo, J.A.M. Chagas, J.F. Mano and R.L. Reis, *Composites Science and Technology*, 2010, **70**, 16, 2310.
38. J. Wang, W. Wu, W. Wang and J. Zhang, *Journal of Applied Polymer Science*, 2011, **121**, 2, 681.
39. M. Rayung, N.A. Ibrahim, N. Zainuddin, W.Z. Saad, N.I.A. Razak and B.W. Chieng, *International Journal of Molecular Sciences*, 2014, **15**, 8, 14728.
40. G. Chandrasekara, M.K. Mahanama, D.G. Edirisinghe and L. Karunanayake, *Journal of the National Science Foundation of Sri Lanka*, 2011, **39**, 3, 243.
41. S. Dasgupta, S.L. Agrawal, S. Bandyopadhyay, S. Chakraborty, R. Mukhopadhyay, R.K. Malkani and S.C. Ameta, *Polymer Testing*, 2008, **27**, 3, 277.
42. H. Ismail, A. Rusli, A.R. Azura and Z. Ahmad, *Journal of Reinforced Plastics and Composites*, 2008, **27**, 16–17, 1877.
43. K.A.J. Dijkhuis in *Recycling of Vulcanized EPDM-Rubber: Mechanistic Studies into the Development of a Continuous Process using Amines as Devulcanization Ads*, University of Twente, Enschede, The Netherlands, 2008. [PhD Thesis]
44. M.A.L. Verbruggen, L. van der Does, J.W.M. Noordermeer, M. van Duin and H.J. Manuel, *Rubber Chemistry and Technology*, 1999, **72**, 4, 731.
45. J.M. Lehn, *Chemistry: A European Journal*, 1999, **5**, 9, 2455.
46. P.A. Brady, R.P. BonarLaw, S.J. Rowan, C.J. Suckling and J.K.M. Sanders, *Chemical Communications*, 1996, **3**, 319.
47. C. Gousse, A. Gandini and P. Hodge, *Macromolecules*, 1998, **31**, 2, 314.
48. C. Toncelli, D.C. De Reus, A.A. Broekhuis and F. Picchioni in *Self-healing at the Nanoscale Mechanisms and Key Concepts of Natural and Artificial Systems*, Eds., V. Amendola and M. Meneghetti, CRC Press, Boca Raton, FL, USA, 2012, p.199.
49. L.M. Polgar, M. van Duin, A.A. Broekhuis and F. Picchioni, *Macromolecules*, 2015, **48**, 19, 7096.
50. K. Chino, M. Ashiura, J. Natori, M. Ikawa and T. Kawazura, *Rubber Chemistry and Technology*, 2002, **75**, 4, 713.
51. K.P. Nair, V. Breedveld and M. Weck, *Macromolecules*, 2008, **41**, 10, 3429.
52. C. Peng and V. Abetz, *Macromolecules*, 2005, **38**, 13, 5575.
53. W.E. Hennink and C.F. van Nostrum, *Advanced Drug Delivery Reviews*, 2002, **54**, 1, 13.
54. M.A.J. van der Mee, R.M.A. l'Abee, G. Portale, J.G.P. Goossens and M. van Duin, *Macromolecules*, 2008, **41**, 14, 5493.
55. M.A.J. van der Mee in *Thermoreversible Cross-linking of Elastomers: A Comparative Study between Ionic Interactions, Hydrogen Bonding and Covalent Cross-links*, Eindhoven University of Technology, Eindhoven, The Netherlands, 2007. [PhD Thesis]
56. P. Cordier, F. Tournilhac, C. Soulie-Ziakovic and L. Leibler, *Nature*, 2008, **451**, 7181, 977.
57. L. Imbernon, E.K. Oikonomou, S. Norvez and L. Leibler, *Polymer Chemistry*, 2015, **6**, 23, 4271.
58. X.N. Chen and E. Ruckenstein, *Journal of Polymer Science, Part A: Polymer Chemistry*, 2000, **38**, 24, 4373.
59. M.A.J. van der Mee, J.G.P. Goossens and M. van Duin, *Polymer*, 2008, **49**, 5, 1239.
60. M.A.J. van der Mee, J.G.P. Goossens and M. Van Duin, *Journal of Polymer Science, Part A: Polymer Chemistry*, 2008, **46**, 5, 1810.

61. R. Gheneim, C. Perez-Berumen and A. Gandini, *Macromolecules*, 2002, **35**, 19, 7246.

62. R. Araya-Hermosilla, A.A. Broekhuis and F. Picchioni, *European Polymer Journal*, 2014, **50**, 127.

63. C. Toncelli, D.C. De Reus, F. Picchioni and A.A. Broekhuis, *Macromolecular Chemistry and Physics*, 2012, **213**, 2, 157.

64. E. Trovatti, T.M. Lacerda, A.J.F. Carvalho and A. Gandini, *Advanced Materials*, 2015, **27**, 13, 2242.

65. L.M. Polgar, E. Hagting, W.J. Koek, M. van Duin and F. Picchioni, *Polymers*, 2017, **9**, 3, 1.

66. L.M. Polgar, M. van Duin and F. Picchioni, *Journal of Visualized Experiments*, 2016, **114**, e54496, 1.

67. L.M. Polgar, R. Blom, J.C. Keizer, M. van Duin and F. Picchioni, *Rubber Chemistry and Technology*, 2017. [In Progress]

68. M.Q. Zhang and M.Z. Rong in *Self-Healing Polymers and Polymer Composites*, Eds., M.Z. Rong and M.Q. Zhang, John Wiley & Sons, Inc, Hoboken, NJ, USA, 2011.

69. S.K. Ghosh in *Self-Healing Materials: Fundamentals, Design Strategies and Applications*, Ed., S.K. Ghosh, Wiley, Weinheim, Germany, 2009, p.1.

70. M.D. Hager and U.S. Schubert in *Self-Healing at the Nanoscale: Mechanisms and Key Concepts of Natural and Artificial Systems*, Eds., M. Amendola and M. Meneghetti, CRC Press, Boca Raton, FL, USA, 2012, p.291.

71. S.D. Bergman and F. Wudl, *Journal of Materials Chemistry*, 2008, **18**, 1, 41.

72. J.A. Syrett, C.R. Becer and D.M. Haddleton, *Polymer Chemistry*, 2010, **1**, 7, 978.

73. D.S. Bag and K.U.B. Rao, *Journal of Polymer Materials*, 2006, **23**, 3, 225.

74. Y. Wang, D.T. Pham and C. Ji, *Coagent Engineering*, 2015, **2**, 1075686.

75. M.R. Kessler, *Proceedings of the Institution of Mechanical Engineers, Part G: Journal of Aerospace Engineering*, 2007, **221**, G4, 479.

76. S. White, N. Sottos, P. Geubelle, J. Moore, M. Kessler, S. Sriram, E. Brown and S. Viswanathan, *Nature*, 2001, **409**, 6822, 794.

77. J.L. Moll, H. Jin, C.L. Mangun, S.R. White and N.R. Sottos, *Composites Science and Technology*, 2013, **79**, 15.

78. T. Yin, M.Z. Rong, M.Q. Zhang and G.C. Yang, *Composites Science and Technology*, 2007, **67**, 2, 201.

79. E. Brown, N. Sottos and S. White, *Experimental Mechanics*, 2002, **42**, 4, 372.

80. X.J. Ye, J. Zhang, Y. Zhu, M.Z. Rong, M.Q. Zhang, Y.X. Song and H. Zhang, *ACS Applied Materials & Interfaces*, 2014, **6**, 5, 3661.

81. H. Klok, E. Rebrov, A. Muzafarov, W. Michelberger and M. Moller, *Journal of Polymer Science, Part B: Polymer Physics*, 1999, **37**, 6, 485.

82. K. Chino and M. Ashiura, *Macromolecules*, 2001, **34**, 26, 9201.

83. C. Peng, A. Gopfert, M. Drechsler and V. Abetz, *Polymers for Advanced Technologies*, 2005, **16**, 11–12, 770.

84. C.X. Sun, M.A.J. van der Mee, J.G.P. Goossens and M. van Duin, *Macromolecules*, 2006, **39**, 9, 3441.

85. F. Maes, D. Montarnal, S. Cantournet, F. Tournilhac, L. Corte and L. Leibler, *Soft Matter*, 2012, **8**, 5, 1681.

86. S.P. Khor, R.J. Varley, S.Z. Shen and Q. Yuan, *Journal of Applied Polymer Science*, 2013, **128**, 6, 3743.

87. C. Wang, N. Liu, R. Allen, J.B-H. Tok, Y. Wu, F. Zhang, Y. Chen and Z. Bao, *Advanced Materials*, 2013, **25**, 40, 5785.

88. Y. Chen, A.M. Kushner, G.A. Williams and Z. Guan, *Nature Chemistry*, 2012, **4**, 6, 467.

89. J. Hentschel, A.M. Kushner, J. Ziller and Z. Guan, *Angewandte Chemie International Edition*, 2012, **51**, 42, 10561.

90. R. Zhang, T. Yan, B. Lechner, K. Schroeter, Y. Liang, B. Li, F. Furtado, P. Sun and K. Saalwaechter, *Macromolecules*, 2013, **46**, 5, 1841.
91. S. Burattini, B.W. Greenland, D.H. Merino, W. Weng, J. Seppala, H.M. Colquhoun, W. Hayes, M.E. Mackay, I.W. Hamley and S.J. Rowan, *Journal of the American Chemical Society*, 2010, **132**, 34, 12051.
92. A. Rekondo, R. Martin, A. Ruiz de Luzuriaga, G. Cabanero, H.J. Grande and I. Odriozola, *Materials Horizons*, 2014, **1**, 2, 237.
93. S. Sontjens, R. Sijbesma, M. van Genderen and E. Meijer, *Journal of the American Chemical Society*, 2000, **122**, 31, 7487.
94. P. Antony and S. De, *Journal of Macromolecular Science, Part C: Polymer Reviews*, 2001, **C41**, 1–2, 41.
95. L.M. Polgar and F. Picchioni in *Non-Covalent Interactions in Synthesis and Design of New Compounds*, Eds., A.M. Maharramov, K.T. Mahmudov, M.N. Kopylovich and A.J.L. Pombeiro, John Wiley & Sons, Hoboken, NJ, USA, 2015, p.431.
96. H. Akimoto, T. Kanazawa, M. Yamada, S. Matsuda, G. Shonaike and A. Murakami, *Journal of Applied Polymer Science*, 2001, **81**, 7, 1712.
97. N. Benetatos and K. Winey, *Journal of Polymer Science, Part B: Polymer Physics*, 2005, **43**, 24, 3549.
98. M. Hara, P. Jar and J. Sauer, *Polymer*, 1991, **32**, 8, 1380.
99. M. Coleman, J. Lee and P. Painer, *Macromolecules*, 1990, **23**, 8, 2339.
100. S. Han, B.H. Gu, K.H. Nam, S.J. Im, S.C. Kim and S.S. Im, *Polymer*, 2007, **48**, 7, 1830.
101. M.E.L. Wouters in *Ionomeric Thermoplastic Elastomers based on Ethylene-Propylene Copolymers*, Eindhoven University of Technology, Eindhoven, The Netherlands, 2000.
102. S.K. Ghosh, P.P. De, D. Khastgir and S.K. De, *Polymer- Plastics Technology and Engineering*, 2000, **39**, 1, 47.
103. L. Li, S. Zhao, C. Chen and Z. Xin, *Journal of Polymer Research*, 2015, **22**, 5, 74.
104. W. Macknight and R. Lundberg, *Rubber Chemistry and Technology*, 1984, **57**, 3, 652.
105. A. Paeglis and F. Oshea, *Rubber Chemistry and Technology*, 1988, **61**, 2, 223.
106. R. Weiss, P. Agarwal and R. Lundberg, *Journal of Applied Polymer Science*, 1984, **29**, 9, 2719.
107. K. Mathew, S. Kumar, A. Lonappan, J. Jacob, T. Kurian, J. Samuel and T. Xavier, *Materials Chemistry and Physics*, 2003, **79**, 2–3, 187.
108. A. Das, A. Sallat, F. Boehme, M. Suckow, D. Basu, S. Wiessner, K.W. Stoeckelhuber, B. Voit and G. Heinrich, *ACS Applied Materials & Interfaces*, 2015, **7**, 37, 20623.
109. J. Bai, H. Li, Z. Shi and J. Yin, *Macromolecules*, 2015, **48**, 11, 3539.
110. T. Scott, A. Schneider, W. Cook and C. Bowman, *Science*, 2005, **308**, 5728, 1615.
111. Y. Amamoto, H. Otsuka, A. Takahara and K. Matyjaszewski, *Advanced Materials*, 2012, **24**, 29, 3975.
112. Y. Amamoto, J. Kamada, H. Otsuka, A. Takahara and K. Matyjaszewski, *Angewandte Chemie International Edition*, 2011, **50**, 7, 1660.
113. J. Canadell, H. Goossens and B. Klumperman, *Macromolecules*, 2011, **44**, 8, 2536.
114. R. Martin, A. Rekondo, J. Echeberria, G. Cabanero, H.J. Grande and I. Odriozola, *Chemical Communications*, 2012, **48**, 66, 8255.
115. M. Pire, S. Norvez, I. Iliopoulos, B. Le Rossignol and L. Leibler, *Polymer*, 2011, **52**, 23, 5243.
116. M. Pire, S. Norvez, I. Iliopoulos, B. Le Rossignol and L. Leibler, *Polymer*, 2010, **51**, 25, 5903.
117. M. Pire, S. Norvez, I. Iliopoulos, B. Le Rossignol and L. Leibler, *Composite Interfaces*, 2014, **21**, 1, 45.
118. M. Pire, C. Lorthioir, E.K. Oikonomou, S. Norvez, I. Iliopoulos, B. Le Rossignol and L. Leibler, *Polymer Chemistry*, 2012, **3**, 4, 946.
119. C. Ding and A.S. Matharu, *ACS Sustainable Chemistry & Engineering*, 2014, **2**, 10, 2217.

120. D. Montarnal, M. Capelot, F. Tournilhac and L. Leibler, *Science*, 2011, **334**, 6058, 965.
121. M. Capelot, D. Montarnal, F. Tournilhac and L. Leibler, *Journal of the American Chemical Society*, 2012, **134**, 18, 7664.
122. M. Capelot, M.M. Unterlass, F. Tournilhac and L. Leibler, *ACS Macro Letters*, 2012, **1**, 7, 789.
123. Y. Lu, F. Tournilhac, L. Leibler and Z. Guan, *Journal of the American Chemical Society*, 2012, **134**, 20, 8424.
124. W. Denissen, G. Rivero, R. Nicolay, L. Leibler, J.M. Winne and F.E. Du Prez, *Advanced Functional Materials*, 2015, **25**, 16, 2451.

2 Environment-sensing rubbers

2.1 Introduction

Smart polymeric rubbers can respond effectively with a considerable change in their properties to small changes in their environment. Environmental stimuli include, for example, mechanical stress, temperature variations and exposure to chemicals.

'Sensing' can be defined as the property of a traditional device to detect events and provide a corresponding output, generally in the form of an electrical signal. An effective sensor provides a fast change of the probe property (i.e., electrical conductivity) under stimuli of its environment by restoring the pristine state in short time and completely recovering the initial energy level. The incorporation of electrically conductive organic or inorganic fillers within a hosting rubber is the most utilised approach for the development of environmentally-sensing materials because they actively combine the probe property with rubber elasticity, low cost and good environmental stability.

Notably, rubbers allow the production of sensor materials with complex shapes that is beneficial for applications in wearable environmental sensors. Electrically conductive rubber composites with high sensitivity have been designed for the preparation of several sensing systems, such as strain sensors, tactile and pressure sensors, gas and liquid sensors, as well as temperature sensors. Moreover, the use of functional nanofillers in the design of nanocomposite sensors has attracted considerable attention due to their high surface area, which endow to the sensor superior sensitivity and reproducibility to various external stimuli and fast response times.

To optimise sensor features it is necessary that the probe property is better preserved (or even improved) by the efficient (nano)dispersion of the functional filler within the rubber matrix. The most sustainable process for the preparation of nanocomposites is that based on the formation of a well-dispersed filler phase during blending with a rubber. The exfoliation of graphitic materials, minerals or pre-formed metal nanostructures during mixing with the polymer allows the dispersion at the micro/nanoscale thanks to the thermomechanical stress induced by the blending process. This procedure is particularly suitable for rubbers and polymers that cannot be processed by solution techniques due to their insolubility in common solvents. High temperature and shear forces in the polymer fluid can break up filler bundles and macro-aggregates, and the high viscosity of the melt prevents their formation during cooling. The melt-blending process allows for the preparation, on a large scale, of filler/polymer mixtures, but sometimes results are less effective than solution blending in terms of disaggregating filler aggregates and agglomerates.

https://doi.org/10.1515/9783110639018-002

In this chapter, the properties and main characteristics of environmental-sensing rubbers are discussed under three main headings: mechanical sensors, temperature sensors and chemical sensors. We address the preparation, inherent properties as well as the methodology of dispersion in the polymer matrix, and also provide relevant examples.

2.2 Mechanical sensors

For the realisation of mechanically stretchable devices and sensors, research on stretchable conductors is required to retain electrical conductivity and mechanical stability under deformation. To this aim, stretchable conductive materials based on soft polymers such as rubbers have attracted enormous interest within the scientific community because rubbers are inherently stretchable, cost-effective, and can even be produced into various types of fabrics. The electrical conductivity of the soft material is endowed by dispersing conductive fillers, and the ultimate conductivity of the composite is governed by the filler content. Notably, there is a volume fraction of conductive filler, known as the 'percolation threshold', at which a continuous interconnecting conductive filler network is created in the composite. Above the percolation threshold, the electrical conductivity is relatively high, whereas below this filler content, the composite acts as an insulator. It has been reported that composites with filler content close to the percolation threshold show the highest conductivity variations (i.e., highest sensitivity) under mechanical solicitations. This result is addressed to the pronounced changes in the relative locations and consequent orientation of the conductive filler along the strain direction of the composite. In conducting filler/polymer composites, applied strain induces filler displacement/sliding on the microscale. These responses give rise to piezoresistive behaviour [1–3]. That is, applied tensile strains result in measurable changes in electrical resistivity across the composite length. Variations in the electrical resistance due to mechanical deformation are evaluated quantitatively by calculating the gauge factor (GF) as $(R - R_0/R_0)/ (1 - l_0/l_0)$ where R is the measured resistance, l is the length of the composite, and R_0 and l_0 their respective initial values. A positive GF denotes a positive piezoresistive effect (i.e., conductive paths are mainly broken down and only few additional paths form with decreasing volume) (Figure 2.1).

By contrast, a negative GF is associated with negative piezoresistive behaviour. This phenomenon is evidenced particularly for fibre-like conducting fillers that are badly aligned within polymers during fabrication, but which acquire perfect anisotropy during strain. Piezoresistivity is usually positive because elongation tends to change the microstructure in such a way that the resistivity becomes larger in the direction of elongation.

Intrinsically conducting polymers such as polypyrrole (Ppy), polythiophene and polyaniline (PANI) have been used due to their high conductivity, good

Progressive strain →

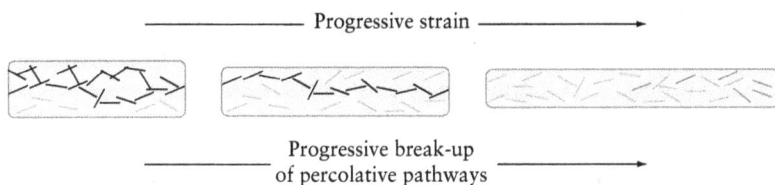

Progressive break-up
of percolative pathways →

Figure 2.1: Mechanical behaviour of conductive filler composite under application of an uniaxial strain (schematic). Percolative pathways are progressively disrupted by the longitudinal force.

stability and ease of synthesis either chemically or electrochemically [4]. Nevertheless, conducting polymer films can endure only limited strain before their rupture. To unravel this issue, they can be embedded within or deposited upon the surface of elastomeric supports and commercial fabrics such as Nylon and polyester [5, 6]. Aw and co-workers proposed to substitute the fabrics with natural rubber (NR) for the preparation of a low-cost, large-strain sensor using Ppy as the conducting polymer. NR has the advantage of combined rigidity and elasticity, which are fundamental characteristics for the preparation of effective strain sensors. The sensor films were prepared by depositing a layer of $FeCl_3$ oxidant onto the pre-stretched NR matrix, which was then exposed to pyrrole vapours to yield a very thin layer (≈2 μm) of conducting polymer. NR pre-stretching at ≤20% of its original length was demonstrated to increase the working range of the strain sensor. The NR surface was previously rendered more hydrophilic *via* plasma activation to ensure deposition of a homogeneous Ppy film without affecting the pristine mechanical features of NR. The Ppy/NR sensor showed the potential to detect large mechanical strains (20%) and negative piezoresistivity with a maximum positive GF of ≈1.9. By contrast, negative piezoresistive behaviour in filler/rubber composites was reported for PANI/acrylonitrile-butadiene rubber (NBR) sheets crosslinked with dicumylperoxide (DCP) [7]. Dodecylbenzenesulfonic acid-doped PANI was blended in a Brabender-type mixer with NBR and then vulcanised with DCP when passed through a two-roll mill. The sheets showed only partial alignment of PANI fibres during processing, but came in intimate contact during external strain. All of these peroxide-vulcanised blends showed good historical memory in terms of their electrical conductivities even after 900 strain-loading and -unloading processes.

The use of anisotropic electrically conductive fillers such as graphite, graphene and carbon nanotubes (CNT) have been used extensively for the realisation of stretchable-resistivity strain sensors for detecting dangerous deformations and vibrations of mechanical parts in many fields of science and engineering [8, 9]. CNT represent the third allotropic form of carbon with unique mechanical, electrical and thermal properties that depend critically on their structural perfection and high aspect ratio (typically >10^2). Single-walled carbon nanotubes (SWCNT) consist of single graphene sheets (monolayer of sp^2-bonded carbon atoms) wrapped into cylindrical tubes with a diameter from 0.7 to 2 nm and lengths up to micrometres.

By contrast, multi-walled carbon nanotubes (MWCNT) consist of concentric assemblies of SWCNT with larger average diameters. Many examples of piezoresistive sensors are referred to rubbers and thermoplastic elastomers (TPE) such as segmented polyurethanes (PU) containing CNT as electrically conductive fillers. The mechanical properties of such PU depend on the physical interactions of the soft polyol and hard diisocyanate segments (Figure 2.2). Highly-conductive thermoplastic polyurethane (TPU) composites have been obtained by blending MWCNT *via* an extrusion process or by ultrasonicating the components in N,N-dimethylacetamide [10, 11].

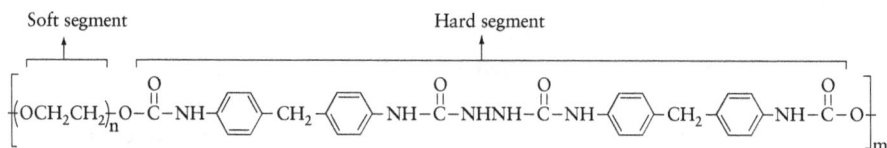

Figure 2.2: Chemical composition of a typical TPU.

Good strain-sensing ability has been achieved with rubber composites already containing 0.015 wt% CNT. Conductive yarns with 2 wt% of CNT content have shown a reproducible strain effect upon cyclic deformations at 30% strain amplitude (Figure 2.3). This characteristic confers the composite rubber good potential as a highly-sensitive mechanical sensor for smart textile applications.

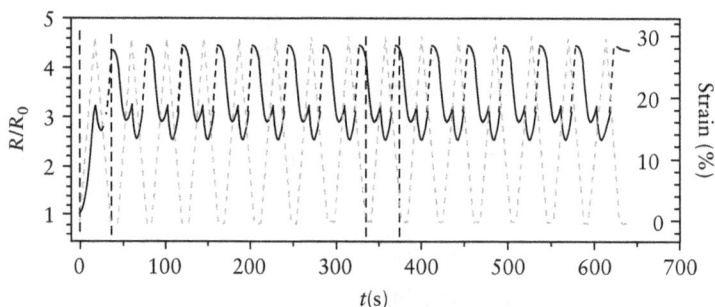

Figure 2.3: Strain-sensing behaviour of coated TPU with 2 wt% of CNT upon cyclic loading at amplitudes of 30% strain. Adapted and redrawn from E. Bilotti, R. Zhang, H. Deng, M. Baxendale and T. Peijs, *Journal of Materials Chemistry*, 2010, **20**, 42, 9449. ©2010, Royal Society of Chemistry [12].

To increase the sensitivity of the TPU stress–strain sensors based on CNT, acid-functionalised MWCNT were proposed. Carboxylic acid functionalities at the surface of MWCNT have been reported to create effective interfacial interactions with the TPU matrix [13]. Through such a simple (but effective) method, strain sensors have been efficiently obtained with large strain-sensing capability (strain≤200%) and a wide range of strain sensitivity with a GF >5.

Very interesting examples have been reported in the literature to detect strains induced on human skin [14]. PU [15], polydimethylsiloxane (PDMS) [16] and NR [16] have been utilised as elastomeric hosts for embedding electrically conductive fillers able to monitor very small strain in skin. Nevertheless, the combination of stretch ability, a high GF and film transparency required for strain sensors for human bodies were difficult to attain. By contrast, an ultrasensitive conductive elastomeric composite based on PU rubber and a stretchable PDMS substrate has been recently proposed by Lee for the realisation of a wearable, transparent, body-attachable strain sensor. The sensor was made of sandwich-like stacked nanocomposite elastomeric films containing acid-functionalised SWCNT and electrically conductive ionomers mixture of poly(3,4-ethylenedioxythiophene) (PEDOT) polystyrene sulfonate (PSS) (Figure 2.4) [14].

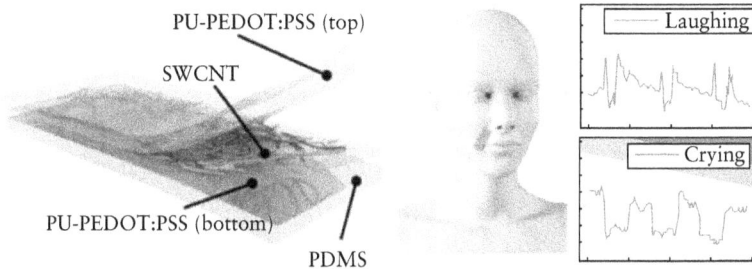

Figure 2.4: Cross-section of a strain sensor consisting of the three- layer stacked nanohybrid structure of PU–PEDOT:PSS/SWCNT/ PU-PEDOT:PSS on a PDMS substrate and a sensor film attached near the mouth. Time-dependent normalised resistance changes are reported as a function of human laughing and crying. Reproduced with permission from E. Roh, B.U. Hwang, D. Kim, B.Y. Kim and N.E. Lee, *ACS Nano*, 2015, **9**, 6, 6252. ©2015, American Chemical Society [14].

The stacked design was realised by spin-coating a solution of PEDOT:PSS stabilised by PU matrix on the PDMS film surface treated by oxygen plasma to favour effective interactions between the layer components. SWCNT was then embedded after PDMS bottom-layer modification with 3-aminopropyltriethoxysilane solution aimed at providing good adhesion with the graphitic material. A third layer of PEDOT:PSS was eventually deposited, thus realising the hybrid sensor film. The film architecture provided by the sandwich-like structure ensured structural and mechanical stability of the composite, enabling the material to undergo towards 1,000 cyclic stretching tests under strains of 20 to 30% without affecting the intrinsic structure. The effective interactions between SWCNT and the conductive PEDOT:PSS matrix allowed use of very low concentrations of the former (1, 3 and 5 mg/mL in the pristine mixture), which contributed to high optical transparency, whereas the modulable content of the latter tuned the sensitivity of the sensor easily. High stretchability of ≤100%, optical transparency of 62%, and a GF of 62 were accomplished, suggesting extremely high strain sensitivity. Notably, when attached to the human face, the film sensor could detect and distinguish between the emotions of laughing and crying (Figure 2.4) as well as eye movements in different directions.

Another example of electrically conducting monodimensional nanostructured (one-dimensional) material utilised in smart sensing is represented by silver nanowires (AgNW). Silver is an attractive metal to examine on the nanoscale due to its extremely high electrical conductivity in the bulk. AgNW with high aspect ratios and diameters <50 nm have attracted considerable interest in the scientific community due to their potential use as interconnects or conductor components in nanodevices [17, 18]. Stretchable conductive rubbers for wearable electronics were prepared by Lee and co-workers using AgNW blended with silver nanoparticles (AgNP) embedded in a polystyrene-*b*-polybutadiene-*b*-polystyrene (SBS) triblock copolymer elastomeric matrix [19]. SBS is a TPE with a block structure investigated primarily as a starting polymer matrix for the preparation of functional materials [20] and used widely in the rubber industry. AgNW with an average diameter and length of about 105 nm and 20 μm, respectively, have been prepared by a typical polyol process using polyvinylpyrrolidone as capping agent [21].

They were embedded in SBS and elastomeric fibres were fabricated *via* a simple wet-spinning method. The AgNP were formed *in situ* on the surface and inner region of AgNW/SBS fibres *via* repeated dipping in an $AgCF_3COO^-$ ethanol solution and successive reduction with hydrazine. Different from the use of AgNP only [22], the presence of an AgNW/AgNP system embedded in the highly stretchable SBS elastomer matrix provided superior initial electrical conductivity (σ_0 = 2,450 S cm^{-1}) and elongation at break of ≤900%. Actually, AgNW behave as conducting bridges between AgNP, thereby preserving most of the composite electrical conductivity even under high strain, with a conductivity loss of ≈4% under 100% strain.

The authors reported strain-sensing behaviour of the fabricated AgNW/AgNP SBS composite fibres. Once integrated into the fingers of a 'smart glove', they can detect human motions effectively (Figure 2.5).

2.3 Temperature sensors

Flexible temperature sensors based on electrically conductive fillers embedded in rubbers have found use because the possibility of tailoring the shape and dimensions of the monitored (human) surface are fundamental issues. For example, monitoring of skin temperature is essential for many applications, especially in sport and medicine [23]. Metal or semiconducting fillers experience a change in resistance with temperature according to Equation 2.1, where α is the temperature coefficient expressed in K^{-1}:

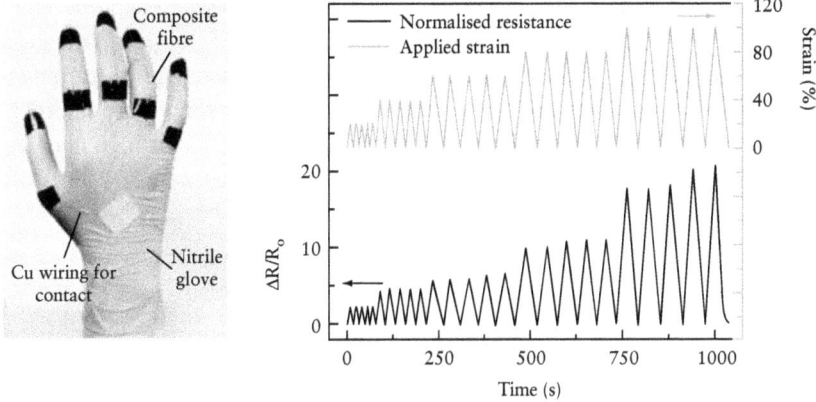

Figure 2.5: A smart glove attached to the composite fibre on each finger and the resistance changes (black lines) as a function of the applied strains (grey lines). Adapted from E. Roh, B.U. Hwang, D. Kim, B.Y. Kim and N.E. Lee, *ACS Nano*, 2015, **9**, 6, 6252 [14].

$$R(T) = R(T_0) \times [1 + \alpha \times (T - T_0)] \tag{2.1}$$

For example, CNT with semiconducting or metallic characters show a resistivity that depends on temperature, which makes CNT/ rubber nanocomposites potentially useful for the fabrication of flexible temperature sensors [24]. Notably, semiconducting CNT show negative $R - R_0/T - T_0$ variations (where R_0 and T_0 are the initial resistance and temperature values, respectively), whereas metallic CNT are characterised by positive dR/dT temperature dependence [25, 26]. The extremely small size and large surface area of exfoliated CNT provide accurate measurements at nanoscale size, thereby allowing very rapid response times. For example, Pucci and co-workers investigated the dispersion of MWCNT in polystyrene- *b*-(ethylene-*co*-butylene)-*b*-styrene (SEBS) mixtures *via* solution processing for the realisation of miniaturised temperature sensors [27]. SEBS is a triblock copolymer consisting of physically crosslinked polystyrene hard domains and hydrogenated polybutadiene soft segments. SEBS was used as a surfactant for MWCNT exfoliation thanks to effective interactions between the styrenic moieties of the polymer and graphitic material. Electrical-resistance measurements were undertaken on flexible films obtained by casting the MWCNT/ polymer dispersions onto a gold electrode. The resulting sensors showed temperature-dependent resistivity with a negative temperature coefficient of $-0.007 \ K^{-1}$, which is comparable with the highest values found in metals ($0.0037-0.006 \ K^{-1}$), but which was partly lost after the first heating cycle up to 55–60 °C. Although the sensors were tested in the temperature range 25–60 °C and this range was far from the lowest reported transition temperatures of the supporting polymer matrix, Pucci and co-workers speculated that the thermal instability of the polymeric matrix may be related to its limited reproducibility. Calisi and co-workers proposed to overcome this issue by annealing the rubber composite at 155 °C for 4 h to obtain a stable percolation network [28]. Thermal annealing led to a remarkable rearrangement of the microdomain structure of the polymer, showing a pronounced improvement in sensor reproducibility [standard deviation (SD) <1%] and very high linearity (R^2 = 0.999) (Figure 2.6).

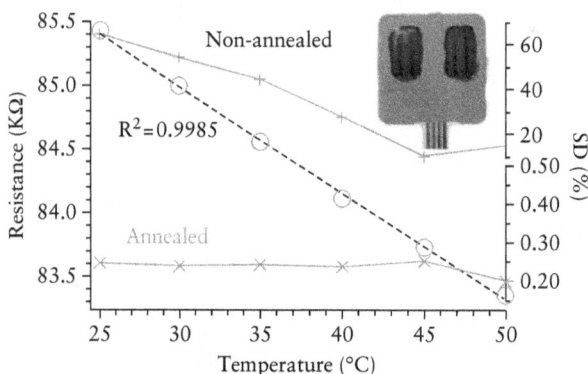

Figure 2.6: Resistance variation and SD of MWCNT/SEBS flexible films as a function of temperature variation in the range 25–50 °C for thermally-annealed and non-annealed sensors. Reproduced with permission from N. Calisi, P. Salvo, B. Melai, C. Paoletti, A. Pucci and F. Di Francesco, *Materials Chemistry and Physics*, 2017, **186**, 456. ©2017, Elsevier [28].

An interesting example was proposed by Park and co-workers for the preparation of rubber composites having a zero temperature coefficient of resistance, a feature that is required for the accurate control of the temperature in heating elements and sensor applications [29]. This feature was accomplished with the preparation of a hybrid bilayer system composed of PDMS layers containing, respectively, MWCNT and carbon black (CB). The layers were prepared by mixing the components in a three-roll mixer and then cured together to fabricate the bilayer composites film. More specifically, the layer based on CB was characterised by a positive temperature coefficient of resistance because the resistivity of the composites increased progressively with temperature due to the difference in the thermal expansion of the polymer and conducting filler. By contrast, the PDMS layer containing MWCNT (12 wt%) showed a negative temperature coefficient of resistance because, in composites containing fillers with high aspect ratios, the interconnection contacts between MWCNT prevailed during thermal expansion. Notably, this peculiar feature conferred the hybrid flexible system a zero temperature coefficient of resistance during heating up to 200 °C.

Metal derivatives based on titanium and bismuth have been also utilised in rubbers as electrically conducting fillers for the preparation of smart materials in shape-memory and electromagnetic-interference shielding devices and temperature sensors [30, 31]. For example, temperature sensors based on styrene-butadiene rubber (SBR) and titanium carbide (TiC) as conducting filler were proposed by Sung [32]. TiC shows a high melting point and electrical conductivity, relatively low coefficient, of thermal expansion, and is a non-polar nature-like graphitic material. The blends were prepared in a two-roll mill for a mixing time of 5 h at 50 °C to ensure uniform dispersion of the TiC particles in the elastomer SBR matrix. The cross-linking of the rubber composite were carried out in press at 150 °C and pressure 500 kPa by using titanium(IV) 2-ethylhexoxide as a curing agent. The percolation threshold was reached for a very low content of TiC (e.g., 0.345 wt%). Two main properties of the rubber composite were found to be dependent on temperature variations between 30 and 120 °C. Actually, thermoelectric power and the dielectric constant were found to increase linearly with temperature and to depend on TiC content.

2.4 Chemical sensors

As soon as a conductive filler/rubber composite experiences external solicitations such as chemical contaminations in the form of liquids or vapours, the conductive network will deform, thus inducing a change in the overall resistivity. This variation in the network relies on the perturbation of the conductive pathways closely associated with a modification of interparticle distances. This feature confers the conductive rubber composites the potential to be designed as chemical sensors. For example, Lu and colleagues prepared NR composites with interconnected graphene-based conductive networks and cellulose nanocrystals (CNC) [33]. CNC have been demonstrated to provide a template/support for graphene oxide to form, after reduction to graphene by hydrazine hydrate, an effective assembled network with a low percolation threshold (0.66 vol%). Notably, the graphene@CNC nanohybrids selectively located in the interstitial space between the NR latex microspheres thus endowed the NR composites with a remarkable resistivity response to organic liquids, especially toluene and chloroform (Figure 2.7).

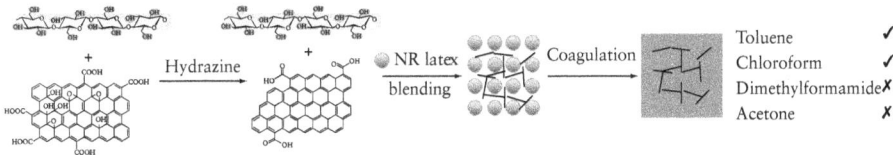

Figure 2.7: Schematic illustration of graphene@CNC NR nanohybrids and sensor responsiveness (toluene and chloroform = responsive; dimethylformamide and acetone = non-responsive) towards the different type of solvents used.

Lu and colleagues addressed the different resistivity changes upon solvent exposures to the different solubility parameters between the NR matrix and targeted organic solvents.

However, the detection of volatile organic compounds (VOC) appears a more appealing issue than liquid sensing. Indeed, VOC are released continuously into the environment and some of them have adverse effects on human health [34, 35].

For example, Zhu and co-workers have recently developed a novel kind of chemical sensor based on a MWCNT/TPU-flexible composite for the determination of VOC [36]. The elastomeric- TPU multifilament produced by melt spinning was immersed into the MWCNT dispersion in chloroform under sonication. The swelling of TPU in chloroform promoted the effective adhesion of MWCNT on the multifilament surfaces because polymer shrinking during composite drying favoured MWCNT phase stabilisation. The obtained MWCNT/TPU composites exhibited rapid and reproducible electrical resistance variations upon successive exposure to diluted VOC [benzene, toluene, $CHCl_3$, tetrahydrofuran (THF), EtOH, acetone and MeOH vapours] and pure air. The composite swelling during VOC exposure caused the disruption of MWCNT percolative pathways with the consequent positive vapour-resistive behaviour.

VOC responses were found to depend on the MWCNT content, vapour concentration, and the solubility parameter of the investigated solvents. For example, MWCNT loadings close to the percolation threshold (0.8 wt%) favoured much larger resistance variations (approximately 900%) when exposed to $CHCl_3$ vapour concentration of 7.0 vol%.

Another interesting example has been recently proposed by Omastová and Mičušík who proposed VOC sensors for monitoring industrial environments [37]. Elastomeric-SBR composites doped by two types of graphitic nanofillers such as CB and MWCNT were prepared by melt mixing and vulcanisation at 150 °C and 20 MPa for no more than 50 min. The percolation threshold was reached at 3.2 wt% of MWCNT and 7.4 wt% of CB, respectively, with the high aspect ratio of the former (>150) being able to create effective percolation pathways within SBR at a much lower loading level. Swelling/ deswelling experiments were performed on sensor composites with compositions slightly above the percolation threshold by using various solvent vapours (acetone, toluene, THF and n-hexane) according to the different Hansen solubility parameter. As already reported earlier, the sensing behaviour is founded on polymer matrix swelling during VOC exposure. The crosslinked network was designed because it stabilises the sensor upon successive cycles of VOC exposure. It was reported that toluene and THF are able to better swell the composite compared to the worse response of n-hexane and acetone due to the missing both polar and H-bonding components of the former and due to the high polarity of the latter (i.e., toluene > THF > n-hexane > acetone). Therefore, acetone was not considered for the sensing investigations. Moreover, MWCNT/ SBR-crosslinked composites were characterised by more restricted mobile polymer chains due to strong bonding between rubber and MWCNT, thus limiting material swelling during VOC exposure and uptake. As a matter of fact, CB/SBR composites reacted much faster to the presence of the investigated VOC with respect to the composites containing MWCNT as conductive fillers. Notably, for VOC with an effective affinity with SBR, the sensor response (positive vapour-resistive behaviour) continued even after the gas exposure is completed due to the presence of an excess of gas uptake that continues its diffusion and polymer swelling.

2.5 Conclusions

This chapter reports the most significant advances in the realisation of rubber composites potentially exploitable as sensors, thanks to the effective dispersion of organic or inorganic electrically-conducting fillers, even with nanostructured characteristics. The possibility of controlling the morphology of elastomeric-based composites has been successfully used to impart polymer matrices new conductive features, which depend on the characteristics of polymers and substrates. The elastomeric nature ensures the realisation of sensors for wearable electronics, as well as highly responsive devices thanks to the materials flexibility and potential swellability. The cross-linked network allows the sensor matrix to be stable upon successive solicitations, thus enabling industrial applications. Intrinsically-conducting polymers, metal (nano) particles and graphitic carbons have been reported to be successfully incorporated within elastomers for the development of sensors to stress–strain deformations, temperature stress and chemical solicitations of liquids and vapours. Notably, the fillers have been embedded in polymer matrices by using different procedures ranging from solution methodologies to melt mixing to investigate the possibility to impart sensing features. All sensing mechanisms are founded on the modification of the percolative networks created by the filler, once embedded in the polymer matrix. Positive or negative resistance variations have been accomplished during the environment solicitation and caused by the progressive rupture or consolidation of the percolative pathways, respectively. Simple or more ingegnerised solutions have been proposed, also with the help of nanotechnology. The use of nanostructured fillers while on one hand amplify the sensor response due to their inherent features and large surface area that increases the interaction with both the rubber matrix and the target, on the other increases the costs and renders their scalable- controlled dispersions still challenging. If these drawbacks are properly addresses, nanostructured electrically-conducting rubbers will have a striking impact in the future of sensors.

References

1. W. Zhifeng and Y. Xiongying, *Nanotechnology*, 2014, **25**, 28, 285502.
2. S. Stassi, V. Cauda, G. Canavese and F.C. Pirri, *Sensors*, 2014, **14**, 3, 5296.

3.   G. Li, L. Wang, C. Leung, R. Hu, X. Zhao, B. Yan and J. Zhou, *RSC Advances*, 2015, **5**, 86, 70229.
4.   J.D. Stenger-Smith, *Progress in Polymer Science*, 1998, **23**, 57.
5.   Y. Li, X.Y. Cheng, M.Y. Leung, J. Tsang, X.M. Tao and M.C.W. Yuen, *Synthetic Metals*, 2005, **155**, 1, 89.
6.   J. Wu, D. Zhou, C.O. Too and G.G. Wallace, *Synthetic Metals*, 2005, **155**, 3, 698.
7.   K.C. Yong, *Journal of Applied Polymer Science*, 2012, **124**, 729.
8.   L. Chen, G.H. Chen and L. Lu, *Nanotechnology*, 2007, **18**, 485202.
9.   J.N. Boland, C.S. Khan, U. Ryan, G. Barwich, S. Charifou, R. Harvey, A. Backes, C. Li, Z. Ferreira, M.S. Mobius, E. Young and R.J. Coleman, *Science*, 2016, **354**, 1257.
10. R. Zhang, H. Deng, R. Valenca, J. Jin, Q. Fu, E. Bilotti and T. Peijs, *Sensor and Actuators A: Physical*, 2012, **179**, 83.
11. R. Zhang, H. Deng, R. Valenca, J. Jin, Q. Fu, E. Bilotti and T. Peijs, *Composites Science and Technology*, 2013, **74**, 1.
12. E. Bilotti, R. Zhang, H. Deng, M. Baxendale and T. Peijs, *Journal of Materials Chemistry*, 2010, **20**, 42, 9449
13. L. Lin, S. Liu, Q. Zhang, X. Li, M. Ji, M. Deng and Q. Fu, *ACS Applied Materials and Interfaces*, 2013, **5**, 12, 5815.
14. E. Roh, B-U. Hwang, D. Kim, B-Y. Kim and N-E. Lee, *ACS Nano*, 2015, **9**, 6, 6252.
15. P. Slobodian, P. Riha, R. Benlikaya, P. Svoboda and D. Petras, *IEEE Sensors Journal*, 2013, **13**, 10, 4045.
16. L. Cai, L. Song, P. Luan, Q. Zhang, N. Zhang, Q. Gao, D. Zhao, X. Zhang, M. Tu, F. Yang, W. Zhou, Q. Fan, J. Luo, W. Zhou, P.M. Ajayan and S. Xie, *Scientific Reports*, 2013, **3**, 3048.
17. C. Liang, K. Terabe, T. Hasegawa and M. Aono, *Nanotechnology*, 2008, **3**, 660.
18. A.L. Pyayt, B. Wiley, Y. Xia, A. Chen and L. Dalton, *Nature Nanotechnology*, 2008, **3**, 11, 660.
19. S. Lee, S. Shin, S. Lee, J. Seo, J. Lee, S. Son and T. Lee, *Advanced Functional Materials*, 2015, **25**, 21, 3114.
20. M. Faraj, E. Elia, M. Boccia, A. Filpi, A. Pucci and F. Ciardelli, *Journal of Polymer Science, Part A: Polymer Chemistry*, 2011, **49**, 15, 3437.
21. M. Bernabo, A. Pucci, H.H. Ramanitra and G. Ruggeri, *Materials*, 2010, **3**, 1461.
22. M. Park, J. Im, M. Shin, Y. Min, J. Park, H. Cho, S. Park, M-B. Shim, S. Jeon, D-Y.Chung, J. Bae, J. Park, U. Jeong and K. Kim, *Nature Nanotechnology*, 2012, **12**, 803.
23. Z.J. Schlader, S.E. Simmons, S.R. Stannard and T. Mundel, *European Journal of Applied Physiology*, 2011, **111**, 1631.
24. L. Dai, P. Soundarrajan and T. Kim, *Pure and Applied Chemistry*, 2002, **74**, 1753.
25. A. Di Bartolomeo, M. Sarno, F. Giubileo, C. Altavilla, L. Iemmo, S. Paino, F. Bobba, M. Longobardi, A. Scarfato, D. Sannino, A.M. Cucolo and P. Ciambelli, *Journal of Applied Physics*, 2009, **105**, 64518.
26. K.S. Karimov, M.T.S. Chani and F.A. Khalid, *Physica E: Low-Dimensional Systems and Nanostructures*, 2011, **43**, 1701.
27. G. Matzeu, A. Pucci, S. Savi, M. Romanelli, J.M. Schnorr, T.M. Swager, F. Di Francesco and A. Pucci, *European Polymer Journal*, 2013, **49**, 1471.
28. N. Calisi, P. Salvo, B. Melai, C. Paoletti, A. Pucci and F. Di Francesco, *Materials Chemistry and Physics*, 2017, **186**, 456.
29. K. Chu, S-C. Lee, S. Lee, D. Kim, C. Moon and S-H. Park, *Nanoscale*, 2015, **7**, 2, 471.
30. M. Frensemeier, J.S. Kaiser, C.P. Frick, A.S. Schneider, E. Arzt, S. Fertig Ray and E. Kroner, *Advanced Functional Materials*, 2015, **25**, 20, 3013.
31. F. El-Tantawy, N.A. Aal, A.A. Al-Ghamdi and E.H. El-Mossalamy, *Polymer Engineering and Science*, 2009, **49**, 3, 592.

32.   Y.K. Sung and F. El-Tantawy, *Macromolecular Research*, 2002, **10**, 6, 345.

33.   J. Cao, X. Zhang, X. Wu, S. Wang and C. Lu, *Carbohydrate Polymers*, 2016, **140**, 88.

34.   S. Endo, B.I. Escher and K.U. Goss, *Environmental Science and Technology*, 2011, **45**, 5912.

35.   S. Manzetti, E.R. van der Spoel and D. van der Spoel, *Chemical Research in Toxicology*, 2014, **27**, 713.

36.   Q. Fan, Z. Qin, T. Villmow, J. Pionteck, P. Pötschke, Y. Wu, B. Voit and M. Zhu, *Sensors and Actuators B: Chemical*, 2011, **156**, 63.

37.   J. Tabaciarova, J. Krajci, J. Pionteck, U. Reuter, M. Omastova and M. Micusik, *Macromolecular Chemistry and Physics*, 2016, **217**, 1149.

3 Optically active elastomers

3.1 Introduction and basic principles

Elastomers which show a change in optical properties by external stimuli are called 'optically active elastomers'. Optical activity can be divided into two types: geometrical and physical. Subcategories of geometrical optical activity are reflection and refraction, which depend strongly on refractive indices. Notably, geometrical optical activity focusses on the pathways of light. Physical optical activity is considered to be a wave, and phenomena such as diffraction, scattering and polarisation have been described.

The velocity of light in a vacuum is equal to the inverse square root of the product of the electric permittivity and magnetic permeability of a vacuum. Light propagation changes from a vacuum to another medium, so the two components of light (i.e., the perpendicular propagating electric and magnetic oscillations) interact with matter *via* electronic and magnetic polarisation. As a result, light is absorbed by matter or loses velocity. If a photon (which has energy higher than the band gap of a non-metallic material) excites an electron, the electron falls back to the valence band by the emission of a photon (i.e., photo-luminescence). Another consequence of light absorption is the increase in kinetic energy of a material. Loss of velocity of light results in reflection and refraction, which are dependent on the refractive index (RI) of a medium. The RI 'n' of a medium is defined according to Equation 3.1, where 'c' is the speed of light in a vacuum and 'v' is the speed of light in the medium:

$$n = \frac{c}{v} \tag{3.1}$$

Thus, the RI of a medium is defined by the electric permittivity and magnetic permeability of the medium. The occurrence of complete refraction, partial refraction and reflection, or total reflection by a medium depends on the angle of incidence, which is determined by Snell's law (Equation 3.2), where 'θ' is the angle of incidence with respect to the normal. No refraction can occur when $\sin \theta > 1$, in which case total reflection occurs. By 'tuning' the RI of a material, reflectivity and refraction can be tuned:

$$n_1 \sin\theta_1 = n_2 \sin\theta_2 \tag{3.2}$$

Light that is neither reflected nor absorbed by a medium is transmitted. However, internal reflection and refraction may cause the intrinsic transparency of a material to differ. Factors that influence the degree of internal scattering are polarisable substances (i.e., dielectrics), porosity and type of crystallinity. Materials in which light is scattered diffusively but reaches the back of the medium are referred to as 'translucent

https://doi.org/10.1515/9783110639018-003

materials'. If no light reaches the back of the medium, the material is referred to as 'opaque'. For non-luminescent materials, the total amount of incoming light is processed into reflection, refraction and absorption. In short, the final optical properties of a medium are dependent on various factors that determine the final colour, transparency and possibility of luminescence.

The basic principles of optical activity are applied in everyday materials with applications ranging from anti-reflective windows to light-emitting diodes (LED) in displays and pigments as colouring agents and probes. Combining optical active components with elastomers yields tailor composite materials that display synergistic effects generally beneficial for the ultimate properties of the composite. In this way, optically active materials can become mechano-responsive and elastomeric materials can become photo-responsive, i.e., the composite is endowed with the features of smart materials [1–3]. Hence, fabrication of optically active elastomers shifts the field of application into new directions with respect to the single components. The application of optically active elastomers is very broad, so only chromogenic elastomers, tropogenic elastomers and elastomeric lenses are discussed in this chapter.

3.2 Chromogenic elastomers

Elastomers that can change their colour by a change in absorption, reflection or emission are defined as 'chromogenic'. Typical methods to obtain such a property in an elastomer are the incorporation of dyes, colloidal structures or fluorescent materials.

Modification of rhodamine into Rh–OH results in a dye with mechanochomic and photo-chromic properties [4]. Moreover, the double hydroxyl functionality of Rh–OH makes it possible to covalently embed Rh–OH dye in polyurethane (PU) by a condensation reaction of the hydroxyl groups with isocyanate groups. The optimal composition of a Rh–OH-embedded PU has been determined at a molar ratio of 0.4 Rh–OH:38.8 tretraethyleneglycol:8.8 triethanolamine:52 hexamethylene diisocyanate by means of mechanical properties [4]. Repetition of the syntheses of the Rh–OH-embedded films demonstrated consistent mechanical properties. Application of uniaxial mechanical stress to the films resulted in isomerisation of the incorporated Rh–OH, which increased the conjugation of the dye. As a consequence, the colour of the film changed from transparent yellow to translucent red. Increasing the applied stress resulted in a higher degree of isomerisation and, therefore, a more prominent mechanochromic effect. However, application of stress to the Rh–OH PU film caused irreversible mechanical deformation. Ultraviolet (UV) patterning by photo-printing of the films resulted in a colour change as well. Both the mechano- and photo-response of the films were reversible, and reversed isomerisation took place at room temperature after 7 days. The recovery speed was increased when the films were exposed to 100 °C for 10 min [4].

Another example of mechanochromism can be effectively provided by opal structures (i.e., colloidal crystals that show reflection properties in the visible range of the electromagnetic spectrum according to the Bragg's law). Therefore, the incorporation of well-arranged colloidal crystals in elastomer matrices provides an effective mechanochromic response. Two typical strategies for the synthesis of elastomeric opals consist of structuring soft core-shell beads by melt flow or by dispersion deposition, with subsequent annealing [5]. A drawback of the melt-flow strategy is that the films obtained by opal orientation are not fully crystalline, notwithstanding a wide range of crystallinity being allowed. Core-shell beads consisting of a hard polystyrene (PS) core, a poly(methyl methacrylate) (PMMA) interlayer and a polyethyl acrylate (PEA) shell arranged by melt flow with subsequent photo-crosslinking with benzophenone do show opal reflection [5]. As expected, increasing the irradiation time or the amount of crosslinker leads to a material with higher crosslink densities and, as a result, a higher Young's modulus. However, the elastomeric opal did not show full mechanical recovery once deformed at 200% of strain [5]. This observation was ascribed to delayed reconfiguration after elongation due to slow molecular relaxation. Mechanochromic activity was visible and confirmed by ultraviolet-visible (UV-Vis) spectroscopy, as can be seen from the extinction peak that shifted from red reflection to blue reflection upon deformation (Figure 3.1).

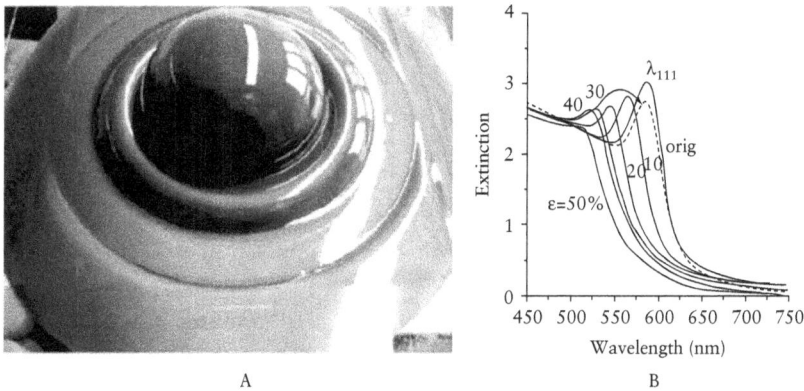

Figure 3.1: A) Visible mechanochromic response to deformation of the elastomeric opal and B) UV-Vis spectrum of a core-shell opal elastomer crosslinked with 2 wt% of benzophenone by 10 min of irradiation; the dotted line shows relaxation into the original shape. Reproduced with permission from B. Viel, T. Ruhl and G.P. Hellmann, *Chemistry of Materials*, 2007, **19**, 23, 5673. ©2007, American Chemical Society [5].

The addition of functional groups at the shell of the core-shell particles results in colloidal crystals that can be crosslinked with different external stimuli. The introduction of hydroxyl functional groups on the shells of the core-shells particles results in colloidal particles that can be crosslinked with diisocyanates at 190 °C

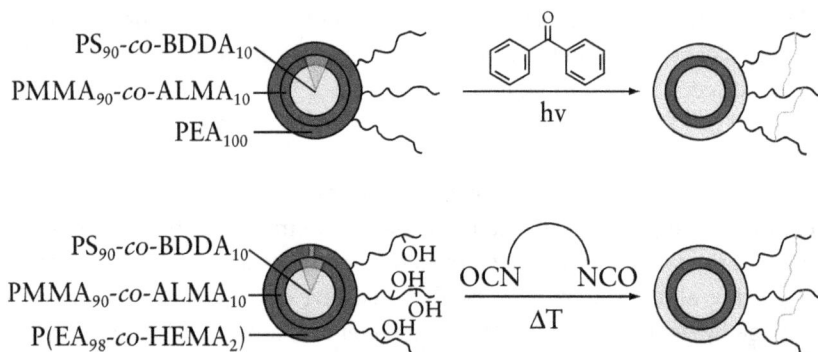

Figure 3.2: Crosslinking processes of core-shell particles (schematic). The top scheme illustrates photo-initiated radical crosslinking with benzophenone. The bottom scheme illustrates thermal-induced crosslinking by the formation of urethane links (ALMA: allyl methacrylate; BDDA: butane-diol diacrylate; HEMA: hydroxyethyl methacrylate; hv: photochemical crosslinking; and ΔT: thermal crossllinking). Reproduced with permission from C.G. Schäfer, B. Viel, G.P. Hellmann, M. Rehahn and M. Gaei, *Applied Materials & Interfaces*, 2013, **5**, 21, 10623. ©2013, American Chemical Society [6].

by the formation of urethane crosslinks [6]. The difference between the composition of this crosslinking method and core-shell particles is shown in Figure 3.2. A comparison between two Crelan oligocyanate crosslinkers (i.e., Crelan® EF403 and Crelan® UI) was made. Differential scanning calorimetry thermograms of the two crosslinkers displayed a transition at a lower temperature for Crelan® EF403, indicating a faster reaction rate of the de-blocking reaction (i.e., activation) of that crosslinker [6]. As a consequence of different de-blocking behaviour, a higher crosslinking efficiency and a higher degree of crosslinking were obtained with the Crelan® EF403 type with respect to the Crelan® UI crosslinker type. Hence, elastomeric opal films display higher Young's moduli and stress at break when crosslinked with Crelan® EF403 crosslinkers. More interesting was the observation during draw–release cycles of the elastomeric opal films. In comparison with the elastomeric opal discussed above, this type of thermo-crosslinked elastomeric opal displayed almost full recovery to its original shape [6]. Moreover, mechanochromic behaviour was observed as a colour shift from red to blue upon elongation. Full recovery of reflection was obtained after the draw–release cycle.

Thus far, two types of approaches have been discussed for the synthesis of chromogenic elastomers: the incorporation of a dye into an elastomeric matrix and the synthesis of elastomeric opals from core-shell beads. A combination of both approaches results in elastomeric opals with triple stimuli-responsive behaviour (i.e., photo-, thermo-, and mechanochromism). Functionalisation of rhodamine B with methacrylamide makes it possible to covalently incorporate the dye into the core or shell of elastomeric opals, depending on the synthetic approach [7]. The resulting elastomeric films can be patterned by UV (i.e., a change in colour of the films triggered by an electromagnetic stimulus). This process is completely thermally reversible.

An intrinsic optically active elastomer is obtained from the polycondensation of 1,10-diisocyanato-4,6-diyne with *bis*(3-aminopropyl)polytetrahydrofuran with subsequent cross-polymerisation (i.e., the formation of ordered crosslinks) by 254-nm UV [8]. The degree of cross-polymerisation was analysed with solid state ^{13}C-nuclear magnetic resonance and was determined to be 10%. The colour of the material changes as a result of cross-polymerisation from colourless into blue due to the progressive formation of the conjugated backbone. Absorption of visible light by the cross-polymerised diacetylene chain is dependent upon the time of exposure to UV light (Figure 3.3).

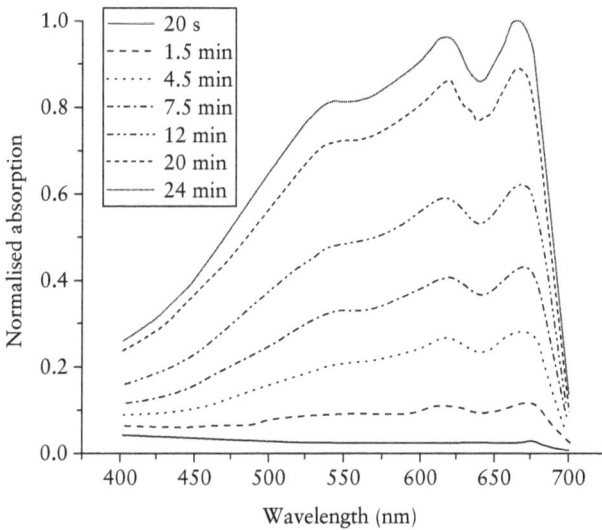

Figure 3.3: Absorption spectra of copoly(ether urea) with a cross-polymerised diacetylene backbone in the visible-light range with increasing time of UV exposure. Adapted from R.A. Koevoets, S. Karthikeyan, P.C.M.M. Magusin, E.W. Meijer and R.P. Sijbesma, *Macromolecules*, 2009, **42**, 7, 2609 [8].

The elongation of copoly(ether urea), which contains cross-polymerised diacetylene, results in a mechanochromic effect (i.e., its colour changes from blue to yellow reversibly up to a strain of 80%). This observation can be ascribed to the fact that a change in the molecular level of the cross-polymerised diacetylene changes the peak absorption and form [8]. Beyond 80% elongation, irreversible deformation takes place in the chromo-active domains of the elastomer, as confirmed by dichroic infrared spectroscopy.

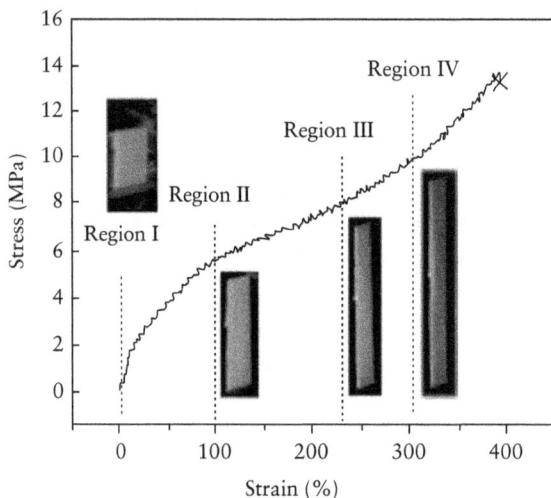

Figure 3.4: Stress–strain curve of solution-cast TPU with 0.5 wt% BBS with images of fluorescence under 365-nm UV. The breakdown of BBS excimers into single molecules by elongation displays a clear shift in fluorescence from cyano-blue to dark blue. Reproduced with permission from S. Bao, J. Li, K.I. Lee, S. Shao, J. Hao, B. Fei and J.H. Xin, *Applied Materials & Interfaces*, 2013, **5**, 11, 4625. ©2013, American Chemical Society [9].

A luminescent elastomer with tuneable luminescence intensity is obtained by the addition of *bis*(benzoxazolyl)stibene (BBS) to a thermoplastic polyurethane (TPU) (Figure 3.4) [9]. The fluorescence of the photo-luminescent elastomer changes from cyano-blue to dark blue upon stretching it to ≤400%. Increasing the amount of BBS in the TPU causes a shift in the emission peak from ≈430 nm for 0.01 wt% to ≈500 nm for 1 wt% with a decreasing pronounced shoulder peak excited by 377-nm UV. The presence of 'side peaks' or shoulders is ascribed to the variation of molecularly dissolved molecules or aggregates of BBS, such as excited dimers (excimers) as well [10]. Post-treatment by heating the specimen to 120 °C results in fully recovered photo-luminescent behaviour and specimen dimensions [9].

Covalent incorporation of a cyano-functionalised hydroxyl-terminated oligo (p-phenylene vinylene) (CN–OPV) in a TPU matrix composed of butanediol, hydroxyl-terminated polytetramethylene glycol and 4,4-methylenebis(phenyl isocyanate) yields a largely reversible mechanochromic elastomer in which the photo-luminescence frequency shifts from an eximer emission at 650 nm to monomer emission at 540 nm when excited by 365-nm UV [11]. After application of stress with subsequent relaxation, the CN–OPV can re-agglomerate through the hard and soft segments of the TPU matrix. Thermal analyses of the CN–OPV TPU elastomer under-pinned this justification. It was observed that an increase in temperature resulted in a shift in photo-luminescence from dominant eximer emission at 20–70 °C to dominant monomer emission at >70 °C [11]. This observation indicates a phase separation of

CN–OPV in TPU induced by kinetic energy. Moreover, a more pronounced mecho-chromic response was observed for samples made by solution polymerisation (Figure 3.5) with respect to samples fabricated by melt polymerisation. Conversely, stress at break was significantly higher for CN–OPV TPU synthesised by melt polymerisation. Both observations indicate that the melt polymerisation

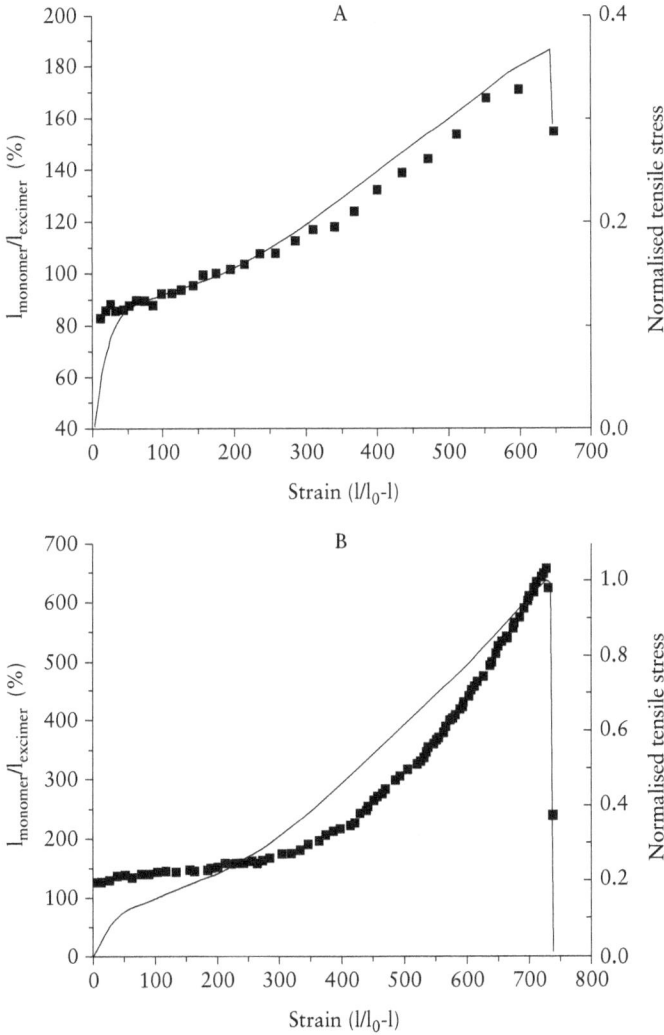

Figure 3.5: Fluorescence intensity of monomer emission over excimer emission and tensile stress, normalised for the maximum tensile stress of B, as a function of strain of two CN–OPV TPU with approximately equimolar compositions. A) Produced by solution polymerisation and B) produced by melt polymerisation. Adapted from B.R. Crenshaw and C. Weder, *Macromolecules*, 2006, **39**, 26, 9581 [11].

synthesis leads to higher-molecular-weight TPU as well as a higher degree of crosslinking. As a result, the soft CN–OPV obtained by solution polymerisation reacts more adequately to deformation by means of mechanochromism than the melt-polymerised sample [11]

Time-based analyses of silicon nanocrystals (SiNC) coated on relaxed and pre-stretched polydimethylsiloxane (PDMS) as well as on a silicon wafer show the effect of oxidation on the photo-luminescence emission peaks by SiNC. An increase in layer thickness of the deposited SiNC on a PDMS substrate results in shift of the emission peak from approximately 770 to 900 nm, excited by a UV/blue LED with peak intensity at 395 nm. Moreover, the sample shows an emission-peak shift towards higher wavelengths as a result of coating PDMS with SiNC in a pre-stretched condition with subsequent relaxation. This behaviour is explained by the fact that the resultant surface wrinkling causes locally increased SiNC domains, which lead to an emission shift towards lower frequencies. This observation is in agreement with the behaviour of SiNC-coated PDMS with increasing deposition thickness. Logically, an increase in pre-stretching during deposition causes emission peaks to shift towards higher wavelengths [12]. The use of an elastomeric substrate in comparison with a dense silicon layer results in an increased oxidation rate of the SiNC-deposited layer. The relatively high permeability of PDMS for gases results not only in oxidation of the surface of SiNC, but oxidation also takes place at the interface layer of PDMS and the SiNC. Hence, the effective layer thickness of the SiNC is reduced faster in comparison with a dense support layer, and a stronger shift in emission frequency is observed [12].

A stress-reporting elastomer can be obtained by filling PDMS with tetrapod-shaped zinc oxide (ZnO). That is, growth of ZnO crystals from the centre of a tetrahedron in the direction of the corners of the tetrahedron. These ZnO structures show green photo-luminescence due to surface irregularity and intrinsically exhibit UV-exciton emission. Therefore, the breakdown of an interconnected ZnO network results in a relative increase in exciton emission [13]. The amount of ZnO tetrapod-shaped particles is critical for the formation of an interconnected network. It was found that the amount of incorporated ZnO tetrapod-shaped particles showed a sufficient mechanochromic effect at 15 wt%. Moreover, the photo-luminescent ratio of green emission over exciton emission was synchronised with the stress–strain curve (Figure 3.6) [13]. It can clearly be seen that an increment in the filler content resulted in a stiffer (but more brittle) elastomeric composite. The fluctuation in photo-luminescent ratio can be ascribed to the relaxation behaviour of the tetrapod-shaped ZnO in the PDMS matrix [13]. This effect was almost absent at a filler concentration of 50 wt%, which indicated that fewer elastic domains were present.

Zinc sulfide (ZnS) is a mechano- and electroluminescent inorganic compound. Therefore, incorporation of ZnS in PDMS matrices results in a smart elastomeric composite that is chromogenic under mechanical or electrical stress. Embedding a ZnS PDMS composite between silver nanowires (AgNW) creates the possibility of applying

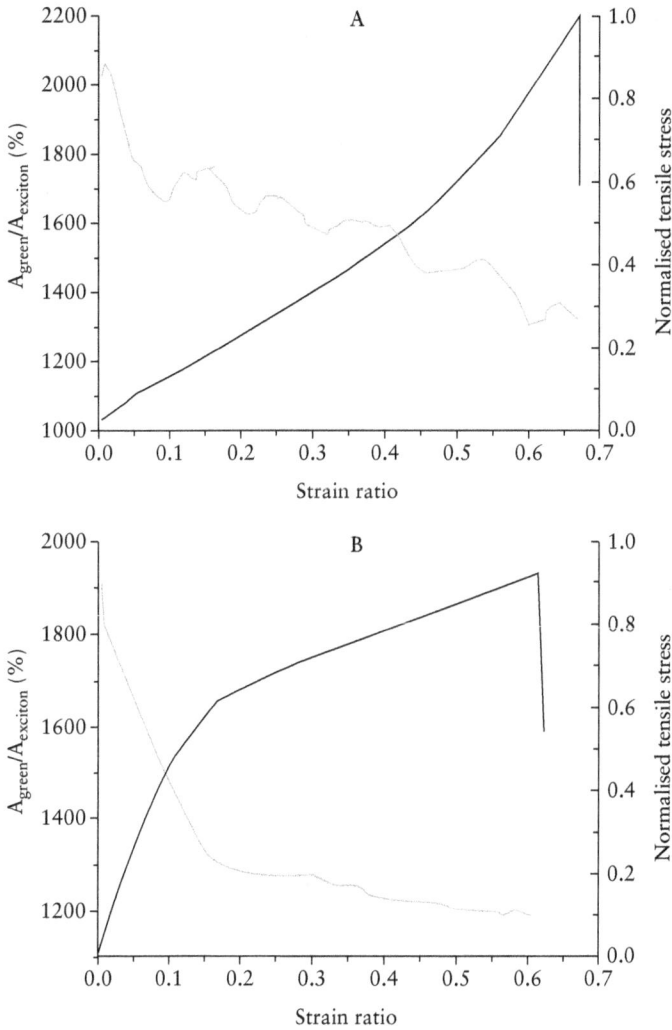

Figure 3.6: Stress–strain curves, normalised for the maximum tensile stress of A, with corresponding emission ratios of green emission over exciton emission, A) PDMS with 15 wt% tetrapod-shaped ZnO and B) PDMS with 50 wt% tetrapod-shaped ZnO. Adapted from X. Jin, M. Goetz, S. Wille, Y.K. Mishra, R. Adelung and C. Zollfrank, *Advanced Materials*, 2013, **25**, 9, 1342 [13].

an electrical field to the composite. ZnS shows green mechanoluminescence (peak emission at 520 nm) and blue electroluminescence (peak emission at 460 nm), so the colour emission of the composite can be tuned by applying mechanical or electrical stress, or applying both stimuli simultaneously. As expected, luminescence intensity is dependent on the stretching releasing rate and frequency of the alternating electrical field. Patterning AgNW results in a functional display [14].

The addition of the colour-converting dye 4-(dicyanomethylene)-2-t-butyl-6-(1,1,7,7-tetramethyljulolidyl-9-enyl)-4H-pyran (DCJTB) to the smart ZnS PDMS elastomeric composite introduces the opportunity to tune the emitted colour. DCJTB absorbs light of a wavelength that precisely overlaps with the mechanoluminescent emission of ZnS, and the absorbed photon is reemitted as red light (peak frequency at 600 nm). Depending on the amount of DCJTB incorporated in the composite, colour emission by mechanoluminescence can be tuned from light green to dark red [15]. By embedding the composite between two electrically conductive components, a chromogenic device that is mechano- and electroactive is produced. However, in comparison with the composite without DCJTB, a wider range of resulting emission can be achieved by varying the dye content [15].

3.3 Tropogenic elastomers

Tropogenic elastomers are smart elastomeric materials that can adapt their geometrical structure by an external stimulus, which causes a change in transparency and reflectivity [16]. In general, a change in transparency also takes place if a chromogenic elastomer changes from transparent to any colour by adaptation in physical or geometrical properties. However, the desired result in the development of chromogenic elastomers is a change in colour, which is accompanied by a change in transparency. Therefore, chromogenic elastomers are discussed only from the viewpoint of a change in colour, and tropogenic elastomers are discussed only from the viewpoint of a change in transparency.

The incorporation of layered double hydroxides (LDH) into a styrene-butadiene rubber (SBR) matrix results in a thermotropic elastomer [16]. Table 3.1 shows the relationship between filler loading and temperature on the degree of thermotropicity, which is defined as the absorbance at 500 nm with respect to the absorbance at 20 °C lower. It can be seen that an increment in filler content from 60 to 100 phr

Table 3.1: Degree of thermotropicity of SBR filled with LDH at different temperatures and filler load.

phr LDH	Temperature (°C) Degree of thermotropicity	
	60	100
4	0	0
60	32	69
100	28	68

Adapted from A. Das, J. Jacob, G.B. Kutlu, A. Leuteritz, D. Wang, S. Rooj, R. Jurk, R. Rajeshbabu, K.W. Stoeckelhuber, V. Galiatsatos and G. Heinrich, *Macromolecular Rapid Communications*, 2012, **33**, 4, 337 [16].

does not result in a further increase in translucency. This thermotropic process is completely reversible. It has been proposed that the change in transparency is caused by thermal rearrangement in the anionic layer of the LDH, which results in a change in RI [16]. As a result, diffusive refraction causes a change from a transparent to a translucent material.

Controlling the superficial structure of an elastomer by mechanical deformation gives rise to tuneable interfacial energy as well as tuneable transparency. Fabrication of PDMS films with a nanopillar-surficial structure shows that the transparent behaviour is tuneable as a function of strain. The films were fabricated by facile template moulding with nanoporous anodic aluminium oxide, whereby the final pillar height could be tuned by changing the depth of anodic aluminium oxide. Subsequently, a microscopic periodic arc-like structure was obtained by stretching the film while it was exposed to UV-ozone radiation [17]. Transmittance of the final mechanotropic PDMS films is shown in Figure 3.7. Interestingly, in the relaxed and strained state, the aspect ratio of the nanopillar arrays did not affect transmittance. A high aspect ratio of the nanopillars apparently contributes to a higher degree of diffusive refraction. However, as a result of microstructural change upon stretching from a periodic arc-like surface to a flat surface, a more pronounced change in transmittance, with respect to the nanostructured-surficial changes, was observed.

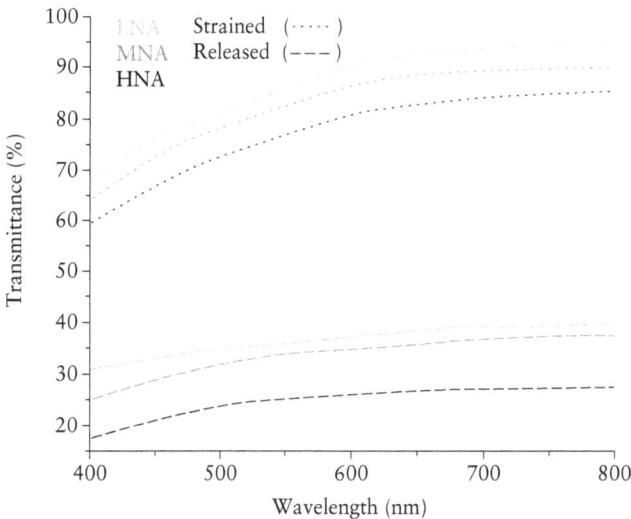

Figure 3.7: Transmittance of the visible spectrum of periodic micro arc-structured PDMS with low-, middle- and high-aspect ratio nanopillar arrays, relaxed and at 30% strain (HNA: high-aspect-ratio nanopillar array; LNA: low-aspect-ratio nanopillar array; and MNA: medium-aspect-ratio nanopillar array). Adapted from S.G. Lee, D.Y. Lee, H.S. Lim, D.H. Lee, S. Lee and K. Cho, *Advanced Materials*, 2010, **22**, 44, 5013 [17].

A reverse approach was used to obtain periodical surface irregularities for diffusive diffraction by applying nanocoating of ZnO onto a highly adhesive acrylic elastomer membrane. In the relaxed state, the composite exhibits a flat surface whereas, in the compressed state, the surface is wrinkled. ZnO was chosen as thin-film coating because it is a highly transparent conductive oxide (≤95%) with respect to other conductive metal oxides. The application of a conduction top layer gives rise to future development of electrically-responsive nanocoated elastomers (e.g., dielectric elastomers). Transmittance of visible light is reduced from ≈90 to ≈10% at a compression of only 5%. Further compression does not result in a significant further decrease in transmittance [18]. Thus, at a compression of 5%, surface wrinkling is already sufficient to diffusively refract visible light.

As discussed above, nanocoating an elastomeric film with metal oxides gives rise to the development of dielectric elastomers with the capability of controlled surface wrinkling by application of an electric field. The design of an electrotropic elastomer results in a smart material in which the surficial structure is controlled by applied voltage (i.e., actuated surface wrinkling). It was found that a smart elastomeric composite consisting of a thin electrode of indium titanium (bottom layer), a dielectric acrylate-based elastomeric (middle layer), and a thin gold electrode (top layer) completely changes its transparency in a reversible manner and changes its reflectivity as a result of surface wrinkling by an applied electric field depending on the angle of incidence. By tuning the thickness of the gold-electrode layer and increasing the electric potential, the broadness and size of the surface wrinkling can be controlled [19].

3.4 Elastomeric lenses

Optical lenses find applications in optical imaging and data storage, where tuneable focus is a certain requirement. The fabrication of stimuli-response elastomeric lenses results in smart materials with tuneable focal lengths. The use of stimuli-responsive elastomeric lenses is preferred over conventional lenses because of their relatively low costs and ease of fabrication. Dip-coating a PDMS film (200–300 µm) with poly (3,4-ethylenedioxythiophene) by immersing PDMS in 3,4-ethylenedioxythiophene (EDOT) with subsequent oxidative polymerisation results in a dielectric elastomer. The transparency of the dielectric elastomer is influenced by the time of oxidative polymerisation (i.e., the transparency is lowered significantly with increasing EDOT polymerisation time). A transparent dielectric elastomeric lens is obtained by fixing the material in a circular lens module with electrodes. It was found that applying a voltage to the lens resulted in a tuneable negative-meniscus lens, which is based on buckling of the dielectric elastomer [20]. The buckling of the film is induced by dielectric compression because the expansive force is translated into a reactive force of the fixed boarders of the lens. A major drawback of this method to make elastomeric actuating lenses is that only lenses with negative focal points can be produced [21]. This problem is overcome by designing fluid-filled dielectric elastomeric lenses. Typically, liquid-filled dielectric elastomeric lenses are constructed by a solid frame with a circular cavity. This cavity is filled with a liquid and the liquid is held in position by two elastomeric membranes to yield a biconvex lens. Actuation is undertaken by dielectric elastomers that are located on the lens [21] or between the lens and frame [22], where a change in focal length is obtained by applying an electric field to the dielectric elastomeric membrane. Due to the strain produced, the curvature of the lenses decreases or increases with subsequent changes in focal length. Electrodes placed on the lens result in flattening actuation [21] and electrodes placed around the lens result in curving actuation [22]. An increase in the initial curvature of the lens results in an increased focal length in the absence of an electric field [21]. It has been argued that lenses with electrodes on their surface should preferably consist of one dielectric elastomer membrane because an additional electrode layer would lower the transparency of the actuating lens [21]. Typically, acrylic elastomers are used as the type of elastomeric membrane [21–23]. Fabrication of an acrylic elastomeric membrane coated with single-walled carbon nanotubes enclosed by silicone oil and passive elastomeric membranes on both sides yields a tuneable lens that can be synthesised not only as a biconvex, but also as a plano-convex and concavo-convex lens [23]. The lens shape is tuned by the liquid pressure within its two cavities. Fabrication of a plano-convex lens with this double-cavity configuration results in a focal-length change of ≤300% if a voltage is applied. At rest, the lens shows a power of ≈67 dpt and, if a voltage of 5,000 V is applied, the focal power decreases to ≈20 dpt [23]. Besides this decrease in focal power, a rapid reversible response is observed. Increasing the focal length (application of 4.5 kV) takes 25 ms, whereas relaxation is possible in 10 ms. This observation is ascribed to the elastic and viscous nature of the acrylic membranes [23].

Astigmatism is an aberration that is found commonly in optical devices. Perhaps the best illustrative example of astigmatism is the human eye, where light does not coincide on the retina if the shape of the cornea or lens is imperfect. A typical example of astigmatism in optical devices is a laser [24]. De-stigmatisation is carried out using cylindrical lenses in which the ratio of the radii of curvature is not equal to 1 along x and y axes. The degree of astigmatism is expressed in astigmatic focal distance (AFD). Fabrication of a microlens consisting of an asymmetrical PDMS lens surrounded by a conductive ring displays actuation behaviour by conductive heating. The PDMS lens expands thermally and the focal length decreases as a result of the Joule effect. Moreover, due to the asymmetry of the lens, the AFD decreases upon heating [24]. Logically, the shape of the focal spots changes 90° with respect to anterior and posterior measurements. A change in AFD from 44 to 0 μm is achieved by applying a current of 30 mA. Besides conductive heating, the astigmatism of lenses can be tuned by other stimuli (e.g., the asymmetrical azimuthal mechanical deformation of elastomeric lenses) [25].

References

1. F. Ciardelli, G. Ruggeri and A. Pucci, *Chemical Society Reviews*, 2013, **42**, 3, 857.
2. A. Pucci and G. Ruggeri, *Journal of Materials Chemistry*, 2011, **21**, 23, 8282.
3. A. Pucci, R. Bizzarri and G. Ruggeri, *Soft Matter*, 2011, **7**, 8, 3689.
4. Z. Wang, Z. Ma, Y. Wang, Z. Xu, Y. Luo, Y. Wei and X. Jia, *Advanced Materials*, 2015, **27**, 41, 6469.
5. B. Viel, T. Ruhl and G.P. Hellmann, *Chemistry of Materials*, 2007, **19**, 23, 5673.
6. C.G. Schaefer, B. Viel, G.P. Hellmann, M. Rehahn and M. Gaei, *ACS Applied Materials & Interfaces*, 2013, **5**, 21, 10623.
7. C.G. Schaefer, M. Gallei, J.T. Zahn, J. Engelhardt, G.P. Hellmann and M. Rehahn, *Chemistry of Materials*, 2013, **25**, 11, 2309.
8. R.A. Koevoets, S. Karthikeyan, P.C.M.M. Magusin, E.W. Meijer and R.P. Sijbesma, *Macromolecules*, 2009, **42**, 7, 2609.
9. S. Bao, J. Li, K.I. Lee, S. Shao, J. Hao, B. Fei and J.H. Xin, *ACS Applied Materials & Interfaces*, 2013, **5**, 11, 4625.
10. J. Birks, I. Munro and D. Dyson, *Proceedings of the Royal Society of London, Series A: Mathematical and Physical Sciences*, 1963, **275**, 1360, 575.
11. B.R. Crenshaw and C. Weder, *Macromolecules*, 2006, **39**, 26, 9581.
12. R. Mandal and R.J. Anthony, *ACS Applied Materials & Interfaces*, 2016, **8**, 51, 35479.
13. X. Jin, M. Goetz, S. Wille, Y.K. Mishra, R. Adelung and C. Zollfrank, *Advanced Materials*, 2013, **25**, 9, 1342.
14. S.M. Jeong, S. Song and H. Kim, *Nano Energy*, 2016, **21**, 154.
15. S.M. Jeong, S. Song, H. Kim, K. Joo and H. Takezoe, *Advanced Functional Materials*, 2016, **26**, 27, 4848.
16. A. Das, J. Jacob, G.B. Kutlu, A. Leuteritz, D. Wang, S. Rooj, R. Jurk, R. Rajeshbabu, K.W. Stoeckelhuber, V. Galiatsatos and G. Heinrich, *Macromolecular Rapid Communications*, 2012, **33**, 4, 337.
17. S.G. Lee, D.Y. Lee, H.S. Lim, D.H. Lee, S. Lee and K. Cho, *Advanced Materials*, 2010, **22**, 44, 5013.
18. M. Shrestha and G. Lau, *Optical Letters*, 2016, **41**, 19, 4433.
19. D. van den Ende, J. Kamminga, A. Boersma, T. Andritsch and P.G. Steeneken, *Advanced Materials*, 2013, **25**, 25, 3438.
20. S. Son, D. Pugal, T. Hwang, H.R. Choi, J.C. Koo, Y. Lee, K. Kim and J. Nam, *Applied Optics*, 2012, **51**, 15, 2987.
21. S. Shian, R.M. Diebold and D.R. Clarke, *Optics Express*, 2013, **21**, 7, 8669.
22. F. Carpi, G. Frediani, S. Turco and D. De Rossi, *Advanced Functional Materials*, 2011, **21**, 21, 4152.
23. S. Shian, R.M. Diebold and D.R. Clarke, *Electroactive Polymer Actuators and Devices*, 2013, **8687**, 86872.
24. S. Lee, W. Chen, H. Tung and W. Fang, *IEEE Photonics Technology Letters*, 2007, **19**, 17, 1383.
25. P. Liebetraut, S. Petsch, J. Liebeskind and H. Zappe, *Light: Science & Applications*, 2013, **2**, 98.

4 Shape-memory elastomers

4.1 Introduction

Shape-memory polymers (SMP) are a class of smart materials. They can recover their 'original shape' if subjected to an external stimulus such as heat, light, pH, solvent exposure, or radiation [1, 2]. Potential applications for such materials have so far been mainly in the biomedical field [3]. This is probably due to the proven combination of shape-memory and drug-release features [4, 5] as well as their extreme sensitivity and response to a given environment [6]. Among all possible stimuli, the application of heat is certainly the most popular, mainly due to the simplicity of the approach. In a typical shape-memory experiment (Figure 4.1), an initial shape is first heated above the glass transition temperature (T_g) of the polymeric material and then a stress is applied to induce a temporary shape [7].

Subsequent cooling below the T_g and removal of the stress freezes the material into a temporary shape. Finally, by heating (without any mechanical stress) above the T_g again, the material can recover its original shape.

When extending this definition to elastomeric materials [shape-memory elastomers (SME)], a conceptual paradox arises. Commercially available rubber products (e.g., tyres) classically comprise low-T_g (typically below room temperature) polymeric chains chemically crosslinked through the use of sulfur or peroxides. In this sense, the material does not display a Tg anymore because the presence of the three-dimensional irreversible network is mainly responsible for the thermal and mechanical properties (i.e., elastic behaviour). This is in stark contrast with the main morphological requirement for SMP, namely the presence of two phases: a 'fixed' and a 'reversible' one [1, 8].

The fixed phase is the result of crystals, entanglements, hydrogen bonding or any type of interaction that can freeze the polymeric chains, thus hindering their relaxation, for example, upon cooling (step 2 in Figure 4.1). The reversible phase, is the one responsible for the switch from the two shapes as it responds to the external stimulus. van der Waals interactions as well as other reversible interactions lead to inter-macromolecular association. The main consequence of this paradox is very important because it immediately defines the possibility of preparing SME from thermoplastic rubbers (for which the T_g is still the leading factor in determining thermal and mechanical behaviours) or from slightly crosslinked rubbers (the crosslinked phase acting here as the fixed phase).

For SME, as for SMP in general, two main parameters [8], both derived from experimental measurements, can be used to quantify shape-memory behaviour: the shape retention rate 'R_f' and the shape-recovery rate 'R_r' as defined by Equations 4.1 and 4.2, respectively:

https://doi.org/10.1515/9783110639018-004

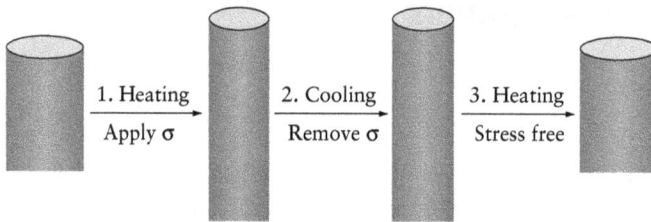

Figure 4.1: General concept of shape memory induced by heat. Adapted and redrawn from J. Li, W.R. Rodgers and T. Xie, *Polymer*, 2011, **52**, 23, 5320 [7].

$$R_f(\%) = \frac{\varepsilon_m}{\varepsilon_u} \times 100 \tag{4.1}$$

$$R_r(\%) = \frac{\varepsilon_u - \varepsilon_p}{\varepsilon_u} \times 100 \tag{4.2}$$

with 'u' being the first deformation (step 1 in Figure 4.1), 'ε_m' the one in the frozen state (step 2 in Figure 4.1) and 'ε_p' the one after the recovery (step 3 in Figure 4.1). The recovery mechanism, as exemplified by the two variables stated above, is more complex than it might seem because usually creep and SME effects might occur simultaneously [9]. These effects are generally more evident for elastomeric materials.

In this chapter, we summarise the state-of-the-art regarding SME by focusing the discussion on the chemical structure of the materials used as well as on their performance as expressed by the two parameters mentioned above. The discussion is divided in terms of the applied external stimulus because this is often the most important when making allowances for the possible applications of SME.

4.2 Heat-induced shape memory in shape-memory elastomers

The application of heat is one of the most popular external stimuli used to switch between the temporary and permanent shape of SME. In view of the discussion above, thermoplastic rubbers, such as styrene-based triblock copolymers (e.g., poly-styrene-*b*-polybutadiene- *b*-polystyrene, and polystyrene-*b*-(ethylene-*co*-butylene)-*b*-styrene (SEBS)), appear to be convenient choices because they display elastomeric behaviour and lack only the presence of a fixed phase hindering chain relaxation in the frozen morphology. Such behaviour can be induced conveniently by simple addition of an extra component in the formulation. In this context, one of the most elegant approaches consists of mixing with a low molecular weight (MW) paraffinic component [10]. This component should be completely miscible with the soft block of the triblock copolymer (SEBS in this case). Upon mixing, the paraffin will locate within the soft ethylene-*co*-butene domains of SEBS (outside the polystyrene blocks), where it crystallises. These crystals then constitute the fixed phase hindering the

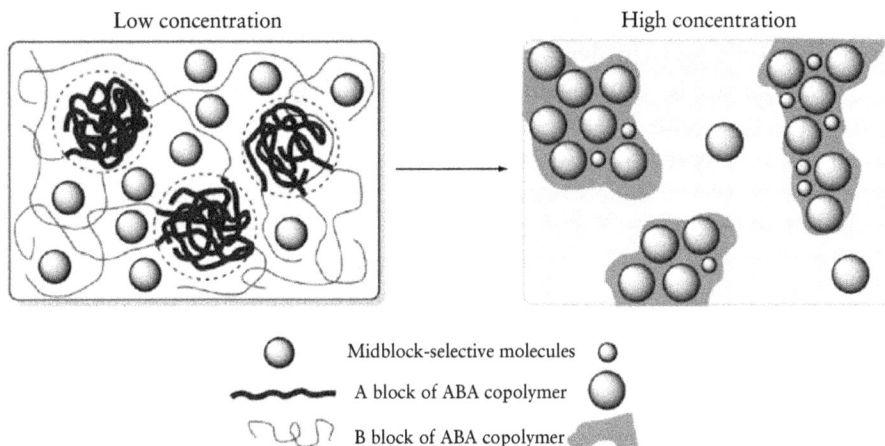

Figure 4.2: Shape-memory effect for SEBS-based materials in the presence of a paraffinic component. A schematic illustration of the microstructure of SME composed of ABA block copolymers and
midblock-selective molecules. Adapted and redrawn from S. Song, J. Feng and P. Wu, *Macromolecular Rapid Communications*, 2011, **32**, 19, 1569 [10].

chain relaxation in the frozen morphology and thus eventually resulting in the desired shape-memory effect (Figure 4.2).

The extra component present here is also crucial in determining the efficiency
of the shape-memory effect as testified by R_f and R_r values (Figure 4.3). In this case,

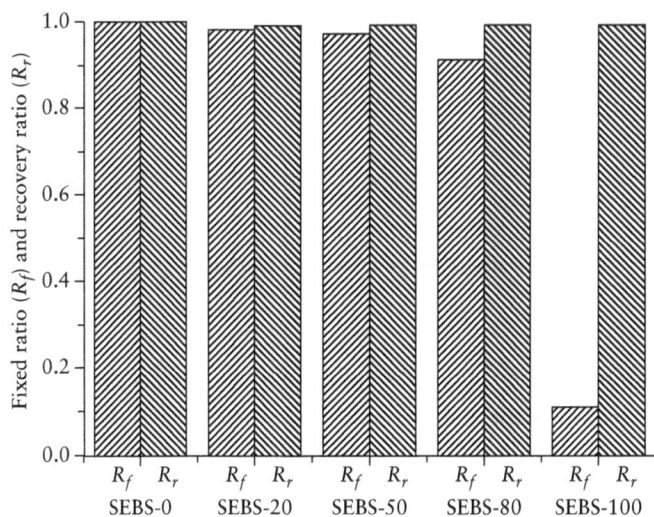

Figure 4.3: Recovery parameters as a function of the composition for SEBS-based SME. The numbers in the coded names indicate the wt% of SEBS in the mixture. Adapted from S. Song, J. Feng
and P. Wu, *Macromolecular Rapid Communications*, 2011, **32**, 19, 1569 [10].

it is clear that a high concentration of paraffin, leading to the formation of a crystal-line matrix as displayed in Figure 4.2, clearly results in more efficient recovery.

For thermoplastic semicrystalline elastomers, addition of such an extra component is obviously not necessary. This conclusion has been reached, among others, for polyglycerol sebacate [11], for which the crystalline domains act as fixed phase while the amorphous regions are the reversible phase. As mentioned above, this effect is not entirely surprising as a slight crosslinked phase may efficiently act as a fixed phase for SME. A series of studies based on natural rubber (NR) have demonstrated this concept clearly [12, 13] while simultaneously showing that application of a mechanical transverse stress might also represent a trigger for shape memory. In this case, such transverse stress can influence the stability of the strain-induced crystals. If crystallinity does not arise spontaneously in the polymer or the addition of a third component is not desired, then a chain extension/coupling approach can be used. For example, the reaction of telechelic end-capped semicrystalline poly-caprolactone (PCL) with epoxidised NR (Figure 4.4) yields the corresponding cross-linked blend [14].

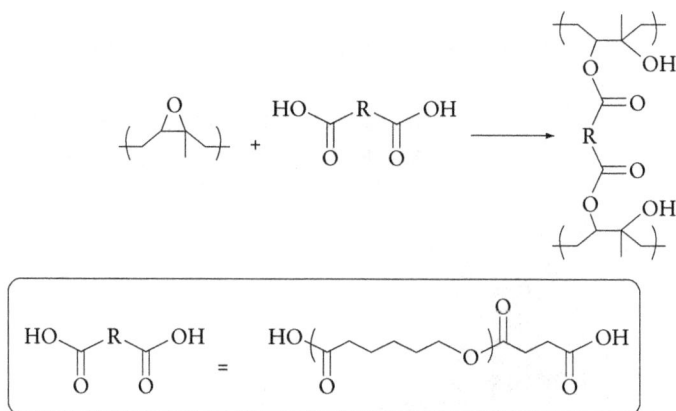

Figure 4.4: Reaction scheme for the crosslinking of NR with telechelic end-capped semicrystalline PCL [14].

Because the two polymers are immiscible, the final morphology consists of a semi-crystalline PCL phase dispersed in an amorphous phase of epoxidised NR. In this case, the shape-memory behaviour is also a function of the composition as well as of the degree of crosslinking.

Simple extension of this concept suggests that hard domains that phase-sepa-rate from an amorphous matrix may also be enough to induce shape-memory prop-erties. Attempts to use this strategy on commercial polymers have been reported recently. In particular, use of non-crosslinked ethylene propylene diene rubber as the soft phase and ethylene–octene copolymers as the hard phase [15] represents a paradigmatic example of a blend composition with relatively large shape-memory efficiencies. Similarly, several studies based on polyurethanes (PU) have confirmed this hypothesis [16, 17]. The approach (e.g., chain extension of a polyisoprene with urethane hard segments [16]) used is elegant when considering possible industrial applications. However, the observed R_r value (85%) also clearly indicates that the hindering effect on the chain relaxation induced by hard domains is probably not as efficient as in crystals. Nevertheless, this effect can be counterbalanced efficiently by crosslinking the system slightly, which can be achieved easily by relying on the chemistry of urethane formation. A cyanate ester of bisphenol A [17] can be modified easily with carboxyl-terminated nitrile rubber chains (Figure 4.5) to yield linear, but also branched, chains through triazine formation.

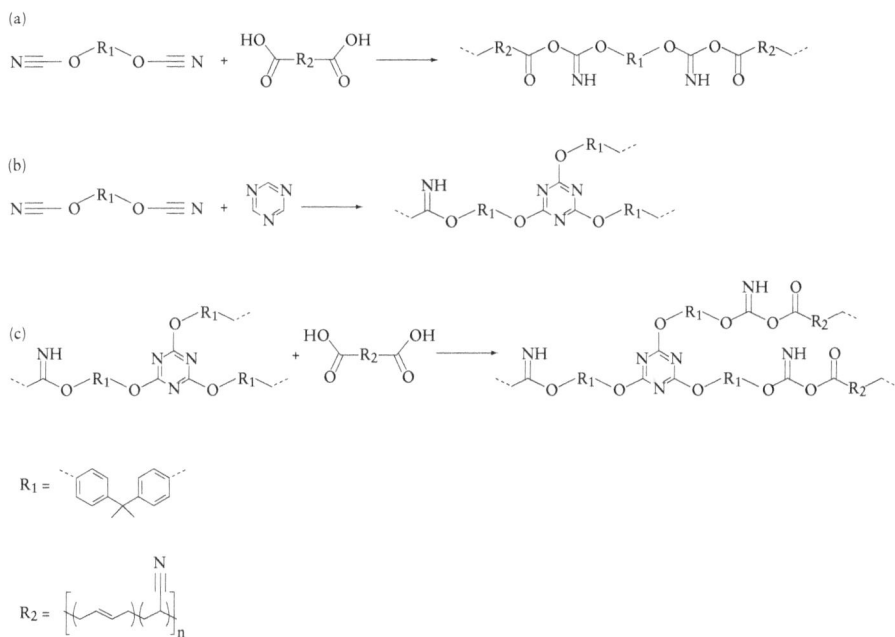

Figure 4.5: Schematic synthetic strategy towards linear and branched/crosslinked SME through urethane chemistry [17].

In this case, the synthetic approach allows fine-tuning of the T_g and has been used to achieve R_f and R_r values close to 100%. The same approach can be used to crosslink epoxidised NR (Figure 4.6) [18]. Also, in this case, R_f and R_r values close to 100% have been reported together with easy modulation of the T_g values as a function of the reaction parameters.

ATA

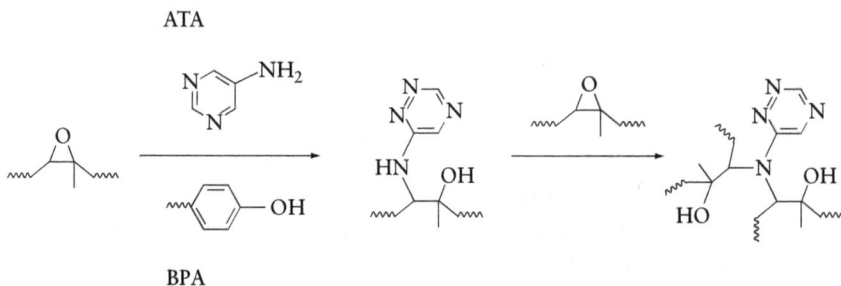

BPA

Figure 4.6: Crosslinking of epoxidised NR with a triazine and bisphenol A (catalyst) [18].

As mentioned above, formation of the fixed phase can be induced by several physical interactions. Among all possibilities, considerable attention has been paid to liquid crystalline (LC) phases [19]. This usually requires specific and dedicated synthetic approaches. A paradigmatic example (Figure 4.7) involves the condensation polymerisation of rigid epoxy-terminated monomers with flexible aliphatic di-carboxylic acids [18].

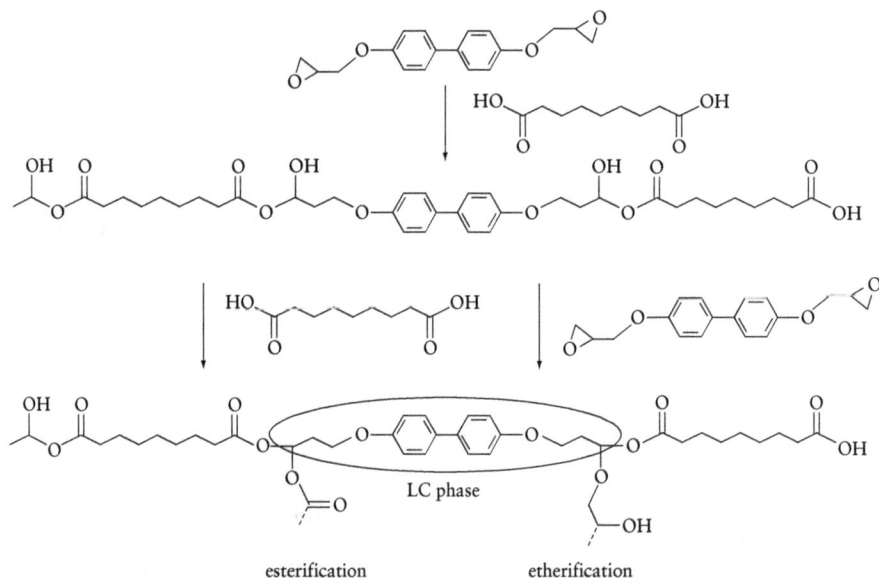

Figure 4.7: LC phases in SME: a possible synthetic approach [18].

Although the preparation might be cumbersome, SME based on LC phases have been shown to exhibit bulk and surface shape-memory effects [20] as well as multiple shape-memory mechanisms [21]. In particular, two-way shape-memory effects (Figure 4.8) have been reported for these materials.

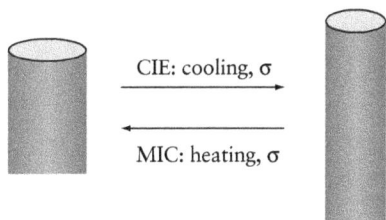

Figure 4.8: Schematic representation of two-way SME after crystallisation-induced elongation (CIE) and melt-induced contraction (MIC). Adapted and redrawn from J. Li, W. Rodgers and T. Xie, *Polymer* 2011, **52**, 23, 5320 [7].

CIE upon cooling and melt-induced contraction (MIC) upon heating are the two basic phenomena that are typically displayed by LC elastomers. These phenomena allow the material to change its shape reversibly as a function of temperature. These phenomena are not solely present in LC elastomers, but have also been reported for slightly crosslinked networks based on ethylene-vinyl acetate copolymers [7] and for systems based on crystalline scaffolds [21].

The same concept of a fixed phase hindering chain relaxation is also easily achievable by means of solid fillers [22, 23]. For poly(L-lactide-*co*-ε-caprolactone) (PLLCA) composites with polyglycolic acid (PGA) microfibres [8], the recovery parameters also have a clear correlation with PGA content. Both increase to ≤95% with increasing PGA microfibre content [8]. In this case, R_f and R_r also clearly increase with the rigidity of the system (i.e., with intake of PGA microfibres). Such clear dependency has been attributed to the second effect of the PGA in the composite: an anti-slippage effect on the PLLCA chains induced by the presence of the solid component.

This concept seems to be quite general and easily applicable because many other works have used the same basic strategy (i.e., through the use of composites) for SME preparation. In particular, the use of small fibres [24] and whiskers [25] seems particularly efficient, even if it might induce clear anisotropic behaviour [26]. Although the use of one-dimensional (1D) fillers is not a strict requirement, it represents a convenient choice because it allows modulation of the end-material properties and tailor them for specific applications. In this context, particular attention has been paid at using specific coatings on the fibres [24] to improve compatibility with the matrix. This would in turn be important in defining the anti-slippage effect (*vide supra*) as well as the mechanical behaviour. The presence of microfibres might also induce faster crystallisation kinetics (if applicable), especially under strain. A study on this phenomenon in SME has not been reported. The use of filler particles instead of fibres (e.g., nano-silica) has also been reported to be effective for NR-based SME [27]. In this case, the presence of the filler is not strictly necessary because the crosslinked phase already acts as the fixed phase in the SME. However, the addition of nano-silica enhances the recovery properties while simultaneously allowing fine-tuning of the mechanical properties.

A relevant advance in this research field has been noted recently [1] and involves the use of thermoplastic fibres into an elastomeric polyanhydride-based

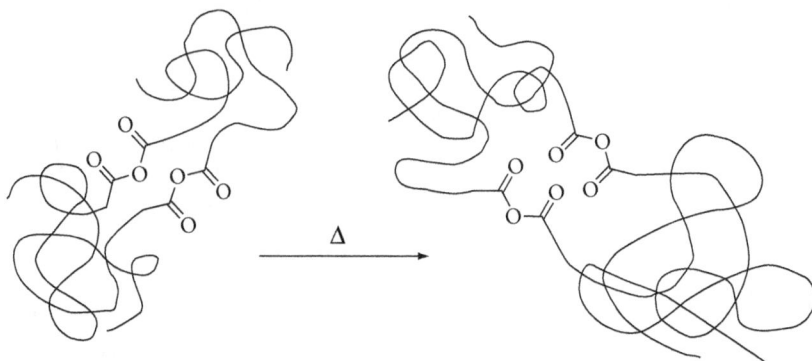

Figure 4.9: Anhydride exchange reactions between different chain segments [1].

material. In this peculiar case, the authors postulated that a dynamic exchange between the anhydride groups of different chains (Figure 4.9) might be responsible
for near-complete reconfiguration of the permanent shape in the solid state.

In addition to the consequences of the observed results for shape-memory behaviour (i.e., the ability to re-programme both shapes easily), it is worth noting
how the proposed concept represents a direct link between SME and intrinsic
self-healing materials. The latter are also based on reversible interactions between the chains that, stimulated by an external stimulus as heat, might be broken
selectively to yielding a covalent (yet adaptable) network [28]. This conceptual synergy has been also exploited earlier [5, 29–32]. It is worth mentioning how this approach might be based on a single (yet chemically modified) elastomer, as in the case
of modified polybutadiene (PB) [29]. Through the use of thiol–ene click reactions (Figure 4.10), two types of modified PB can be prepared readily: those that display amino
groups and acid groups.

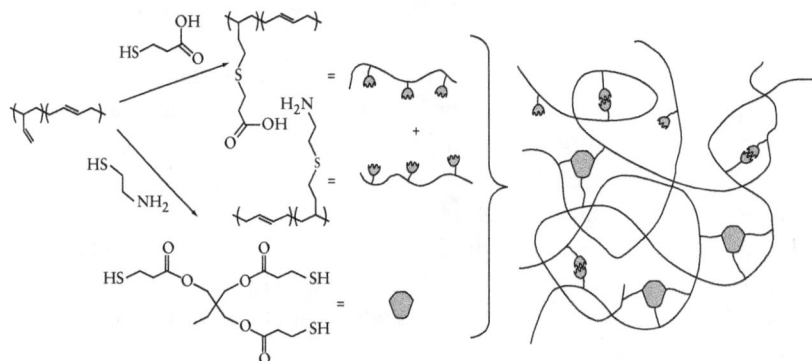

Figure 4.10: Synthetic strategies for the preparation of PB-based SME by self-assembly and
photo-crosslinking [29].

By mixing the two polymers together, one exploits the presence of electrostatic interactions as well as hydrogen bonding (both being thermally reversible) in switching from the two shapes. The material is slightly crosslinked (to generate the fixed phase) and should be self-healing in the proposed conditions as well. The measured R_f and R_r values of >95% clearly testify to the possibility of using thermally-reversible interactions (others than van der Waals forces) for the preparation of smart materials.

4.3 Other external stimuli

Besides heat, other external stimuli can be used for shape recovery. A considerable number of studies have focused on solvent-induced shape-memory effects even if the solvent is present as a vapour [33, 34]. All the reported results can be clarified when making allowances for the plasticising effect of the solvent on the polymeric materials. A relatively simple and elegant theoretical study based on the Flory–Huggins miscibility theory and relying on the use of Hildebrand solubility parameters [35] allows easy prediction of the proposed effect. The free energy of mixing 'W_m' in a system consisting of polymeric chains and solvent is given by Equation 4.3 with 'k' being the Boltzmann constant, 'T' the absolute temperature, 'v' the volume of a solvent molecule, 'C' the number of solvent molecules and 'X' the Flory–Huggins interaction parameter. Simple derivation of W_m as a function of C yields the chemical potential of the polymer 'μ' according to Equation 4.4.

$$W_m = \frac{kT}{v} \times \left[vC \times \log(1 + \frac{1}{vC}) - \frac{X}{1 + vC} \right] \tag{4.3}$$

$$\mu = \frac{\partial W_m}{\partial C} = kT \times \left[-\log(1 + \frac{1}{vC}) + \frac{1}{1 + vC} + \frac{X}{(1 + vC)^2} \right] \tag{4.4}$$

Comparison of this theoretical equation with experimental data based on PU SME [35] yields a final equation (Equation 4.5) that can predict the effect of the volume fraction of the solvent 'ϕ' on the T_g of the polymer. This allows for easy prediction of the effect of the solvent presence on the thermal properties and thus shape-memory behaviour.

$$T_g = 37e^{-0.06\phi^3} - 0.6(1 - e^{-\phi}) + 273 \tag{4.5}$$

The use of water as solvent is particularly interesting for obvious reasons connected to safety and application perspectives. SME based on cellulose nanowhiskers dispersed in a PU matrix display shape-memory effects induced by water [36]. In this case, the concept relies on the disruption (and reformation upon drying) of the percolation network of the nanowhiskers in the presence of the external stimulus. This mechanism does not rely on the plasticising effect and, as such, represents a novel and relatively fast switch method.

Among the wide range of external stimuli, the use of a magnetic field has been receiving some attention. In this case, the often-cumbersome incorporation of magnetic-responsive moieties in the SME is counterbalanced by the simplicity of the system [37]. The basic experimental approach (Figure 4.11) might consist of the final recovery of the shape in the magnetic field.

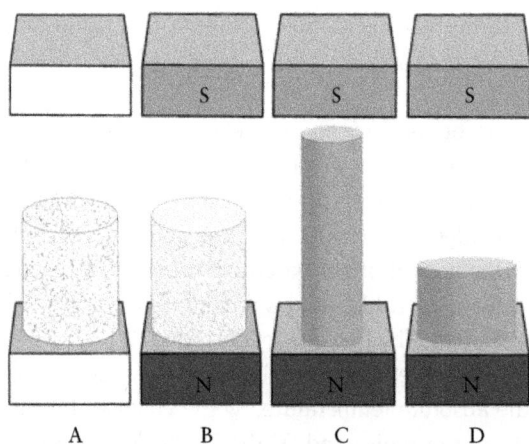

Figure 4.11: Schematic representation of SME (A) in the initial state without a magnetic field, (B) in the initial state in the magnetic field, (C) in a stretched and (D) compressed state in the magnetic field with 'N' the north pole and 'S' the south pole of the magnets used. Adapted and redrawn from L.V. Nikitin, G.V. Stepanov, L.S. Mironova and A.I. Gorbunov, *Journal of Magnetism and Magnetic Materials*, 2004, **272–276**, 2072 [37].

As mentioned above, the observed effect relies on the possibility to disperse magnetic-responsive moieties into a polymeric matrix. A possible approach consists of the dispersion of Fe_3O_4 particles into the polymer [38]. A good example of a material that makes use of this approach was prepared by copolymerisation of 2-methoxyethyl acrylate and N,N-dimethylacrylamide (Figure 4.12).

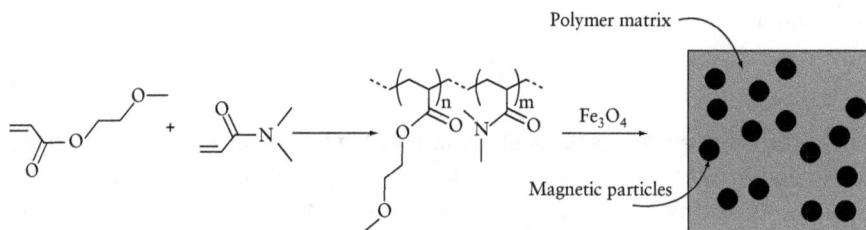

Figure 4.12: Schematic preparation of magnetic SME [38].

The presence of amino groups, although not explicitly mentioned, might be crucial in stabilising the Fe-based particles. One might also notice, with respect to the conceptual analogy between shape memory and self-healing, that the prepared materials display self-healing properties as induced by infrared (IR) light (Figure 4.13). This is not surprising as the molecular basis of both phenomena is identical: the existence of a thermally-reversible network.

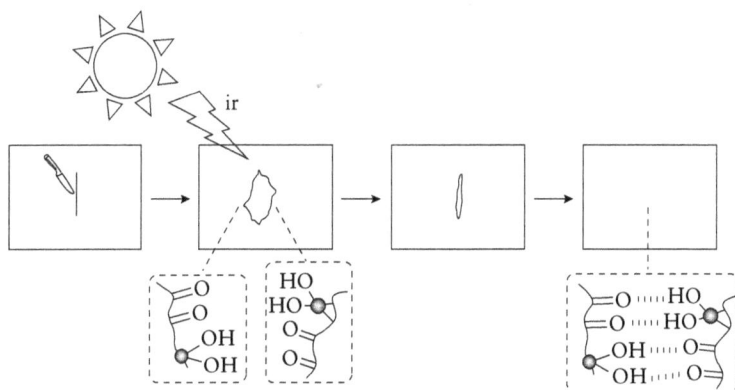

Figure 4.13: Illustrations (a) and real images (b) of shape memory-assisted self-healing. Adapted and redrawn from X.Q. Feng, G.Z. Zhang, Q.M. Bai, H.Y. Jiang, B. Xu and H.J. Li, *Periodical Macromolecular Materials and Engineering*, 2016, **301**, 2, 125 [38].

This synergy in response to different stimuli (and the possibility of multiple responses at all) is very interesting when considering the idea of smart materials. PB-based SME (*vide supra* [29]) have been shown to display shape memory as a result of the application of heat or ultraviolet (UV) light. This is possible because the modified PB contains groups that are responsive to UV light (double bonds) as well as groups that are sensitive to heat (ionic and hydrogen bonding).

The choice of the most convenient stimulus might be dependent on the final application conditions, but also on the used SME. Nanocomposites based on PCL (connected through urethane bonds) and reduced graphene oxide (RGO) [39] can be prepared *via* a relatively simple strategy (Figure 4.14).

The fact that this material can respond to three external stimuli sets it apart from the broad variety of SME. In this case, the RGO provides the material with electroconducting properties because it is embedded in the crosslinked network *via* physical interactions (e.g., hydrogen bonding). The shape-recovery efficiency (Figure 4.15) is clearly dependent on the external stimulus, with heat representing the one with the slowest response.

In analogy to similar trends found for other composites (*vide supra*), the filler intake seems to have a positive influence on the shape- recovery process.

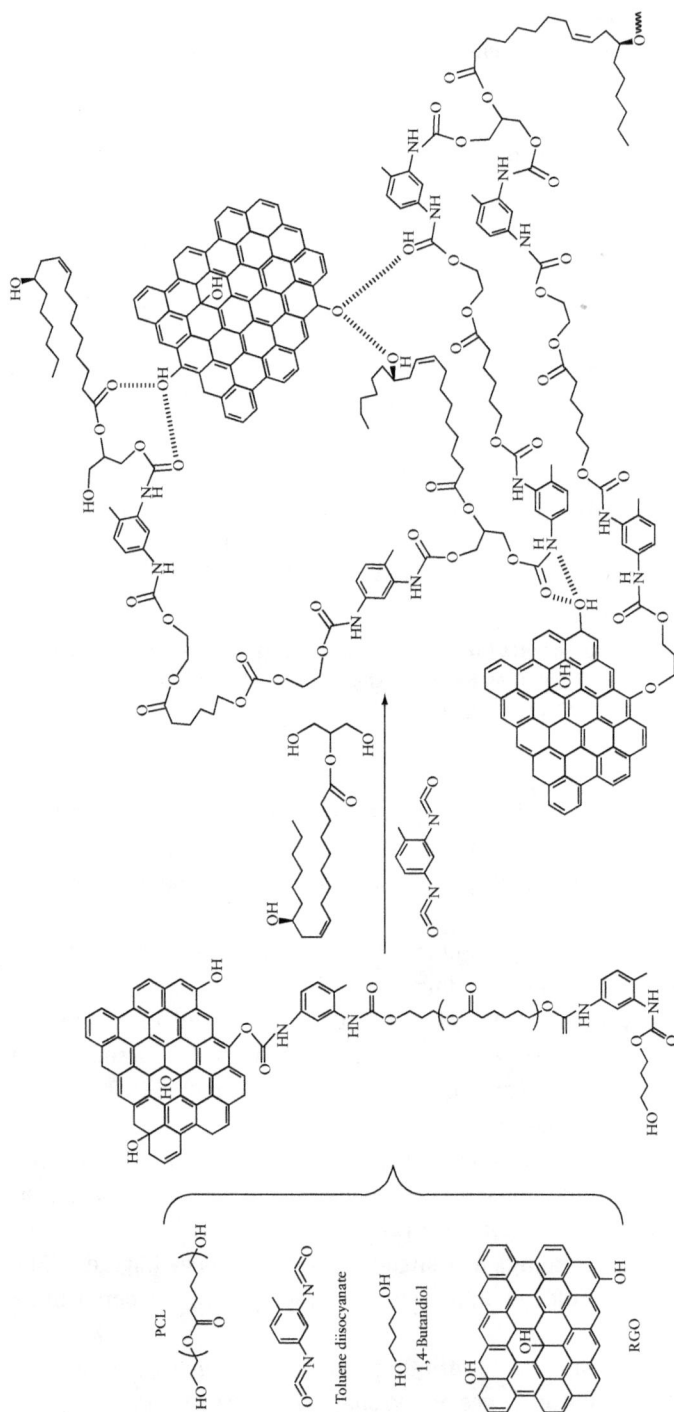

Figure 4.14: Preparation strategy for SME based on nanocomposites RGO [39].

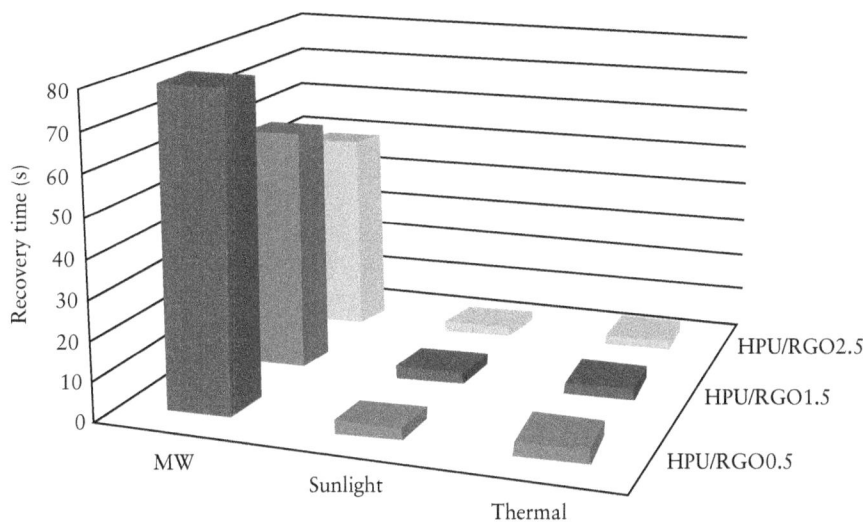

A

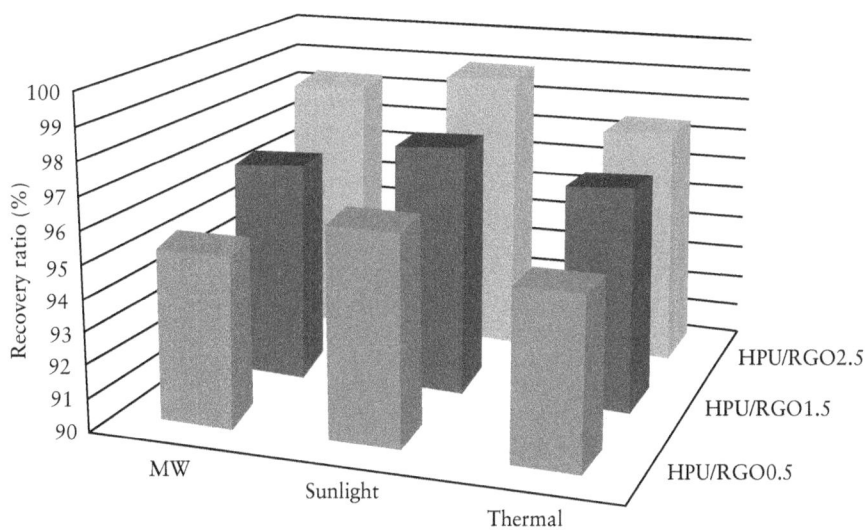

B

Figure 4.15: Bar graph on the recovery time (A) and recovery ratio (B) of the exemplified nanocom-posites based on PCL (connected through urethane bonds), hyperbranched polyurethane (HPU) and RGO. Adapted and redrawn from S. Thakur and N. Karak, *Periodical Journal of Materials Chemistry A*, 2014, **2**, 36, 14867 [39].

References

1. M.I. Lawton, K.R. Tillman, H.S. Mohammed, W. Kuang, D.A. Shipp and P.T. Mather, *ACS Macro Letters*, 2016, **5**, 2, 203.
2. T. Ware, W. Voit and K. Gall, *Radiation Physics and Chemistry*, 2010, **79**, 4, 446.
3. A. Lendlein and R. Langer, *Science*, 2002, **296**, 5573, 1673.
4. N. Coi and A. Lendlein, *Soft Matter*, 2007, **3**, 7, 901.
5. M.C. Serrano, L. Carbajal and G.A. Ameer, *Advanced Materials*, 2011, **23**, 19, 2211.
6. R. Hoeher, T. Raidt, F. Katzenberg and J.C. Tiller, *ACS Applid Materials & Interfaces*, 2016, **8**, 22, 13684.
7. J. Li, W.R. Rodgers and T. Xie, *Polymer*, 2011, **52**, 23, 5320.
8. L. Wang, H. Chen, Z. Xiong, X. Pang and C. Xiong, *Macromolecular Materials and Engineering*, 2010, **295**, 4, 381.
9. X.L. Wu, W.M. Huang and H.X. Tan, *Joural of Polymer Research*, 2013, **20**, 8, 150.
10. S. Song, J. Feng and P. Wu, *Macromolecular Rapid Communications*, 2011, **32**, 19, 1569.
11. W. Cai and L. Liu, *Materials Letters*, 2008, **62**, 14, 2171.
12. B. Heuwers, D. Quitmann, R. Hoeher, F.M. Reinders, S. Tiemeyer, C. Sternemann, M. Tolan, F. Katzenberg and J.C. Tiller, *Macromolecular Rapid Communications*, 2013, **34**, 2, 180.
13. B. Heuwers, D. Quitmann, F. Katzenberg and J.C. Tiller, *Macromolecular Rapid Communications*, 2012, **33**, 18, 1517.
14. Y. Chang, J. Eom, J. Kim, H. Kim and D. Kim, *Journal of Industrial and Engineering Chemistry*, 2010, **16**, 2, 256.
15. T. Chatterjee, P. Dey, G.B. Nando and K. Naskar, *Polymer*, 2015, **78**, 180.
16. X. Ni and X. Sun, *Journal of Applied Polymer Science*, 2006, **100**, 2, 879.
17. Z. Yamei, Z. Doudou and G. Li, *Colloid and Polymer Science*, 2014, **292**, 10, 2707.
18. Y. Li, C. Pruitt, O. Rios, L. Wei, M. Rock, J.K. Keum, A.G. McDonald and M.R. Kessler, *Macromolecules*, 2015, **48**, 9, 2864.
19. K. Hiraoka, N. Tagawa and K. Baba, *Macromolecular Chemistry and Physics*, 2008, **209**, 3, 298.
20. K.A. Burke and P.T. Mather, *Polymer*, 2013, **54**, 11, 2808.
21. A.R. Garcia-Marquez, B. Heinrich, N. Beyer, D. Guillon and B. Donnio, *Macromolecules*, 2014, **47**, 15, 5198.
22. T. Rey, F. Razan, E. Robin, S. Faure, J-L. Cam, G. Chagnon, A. Girard and D. Favier, *International Journal of Adhesion and Adhesives*, 2014, **48**, 67.
23. Q. Ge, X. Luo, E.D. Rodriguez, X. Zhang, P.T. Mather, M.L. Dunn and H.J. Qi, *Journal of the Mechanics and Physics of Solids*, 2012, **60**, 1, 67.
24. Y. Ko, Y. Lee, K. Devarayan, B. Kim, T. Hayashi and I. Kim, *Materials Letters*, 2014, **131**, 128.
25. L. Wang, H. Chen, L. Zang, D. Chen, X. Pang and C. Xiong, *Journal of Polymer Research*, 2011, **18**, 3, 329.
26. E.D. Rodriguez, D.C. Weed and P.T. Mather, *Macromolecular Chemistry and Physics*, 2013, **214**, 11, 1247.
27. T. Lin, S. Ma, Y. Lu and B. Guo, *ACS Applied Materials & Interfaces*, 2014, **6**, 8, 5695.
28. C. Toncelli, D.C. De Reus, A.A. Broekhuis and F. Picchioni in *Self-healing at the Nanoscale Mechanisms and Key Concepts of Natural and Artificial Systems*, Eds., V. Amendola and M. Meneghetti, CRC Press, Boca Raton, FL, USA, 2012, p.199.
29. D. Wang, J. Guo, H. Zhang, B. Cheng, H. Shen, N. Zhao and J. Xu, *Journal of Materials Chemistry A*, 2015, **3**, 24, 12864.
30. T. Tsujimoto, K. Toshimitsu, H. Uyama, S. Takeno and Y. Nakazawa, *Polymer*, 2014, **55**, 25, 6488.

31. I.M. Van Meerbeek, B.C. Mac Murray, J.W. Kim, S.S. Robinson, P.X. Zou, M.N. Silberstein and R.F. Shepherd, *Advanced Materials*, 2016, **28**, 14, 2801.
32. D. Perez-Foullerat, S. Hild, A. Mucke and B. Rieger, *Macromolecular Chemistry and Physics*, 2004, **205**, 3, 374.
33. D. Quitmann, M. Dibolik, F. Katzenberg and J.C. Tiller, *Macromolecular Materials and Engineering*, 2015, **300**, 1, 25.
34. D. Quitmann, N. Gushterov, G. Sadowski, F. Katzenberg and J.C. Tiller, *ACS Applied Materials & Interfaces*, 2013, **5**, 9, 3504.
35. H. Lu and S. Du, *Polymer Chemistry*, 2014, **5**, 4, 1155.
36. Y. Zhu, J. Hu, H. Luo, R.J. Young, L. Deng, S. Zhang, Y. Fanc and G. Ye, *Soft Matter*, 2012, **8**, 8, 2517.
37. L. Nikitin, G. Stepanov, L. Mironova and A. Gorbunov, *Journal of Magnetism and Magnetic Materials*, 2004, **272–276**, 2072.
38. X.Q. Feng, G.Z. Zhang, Q.M. Bai, H.Y. Jiang, B. Xu and H.J. Li, *Macromolecular Materials and Engineering*, 2016, **301**, 2, 125.
39. S. Thakur and N. Karak, *Journal of Materials Chemistry A*, 2014, **2**, 36, 14867.

5 Magnetorheological elastomers

5.1 Background and basic principles

Magnetorheological (MR) materials are viscous or viscoelastic materials (e.g., fluids, gels or elastomers) with suspended magnetisable particles. A typical feature of these materials is that their shape can be controlled by magnetic fields. Magnetorheological fluids (MRF) were first discovered by Rabinow [1] and have been studied extensively in the last two decades. Newtonian behaviour is observed for MRF if a magnetic field is absent. If a MRF is subjected to a magnetic field, the suspended magnetic particles (e.g., iron-based particles) magnetise (i.e., the magnetic dipoles align to the magnetic field) and orientate in the direction of the magnetic field lines in ordered columns. As a result, the MRF switches into solid state in which the yield stress of the MRF is determined by the magnetic-field strength. This change in material properties by means of a modulus and damping is called 'the MR effect' and is completely reversible. A major drawback of MR smart material is the sedimentation of the magnetic particles in the fluid because this influences the performance of the MRF significantly. This problem can be overcome by dispersing magnetisable particles in a solid, elastomeric matrix to yield 'magnetorheological elastomers' (MRE). MRE consist of an electrically-isolating elastomeric matrix, magnetisable fillers, and additives.

Characteristic elastomers that are typically used for the preparation of MRE are natural rubber (NR), silicone rubber, polybutadiene (PB) and polyurethane (PU). Carbonyl iron (highly pure iron particles) is commonly used as magnetisable filler, and silicone oil may be used as an additive to increase matrix–filler interactions [2–4]. The final properties of MRE are highly dependent on the type of matrix, size and alignment of the magnetic particles, and the crosslink density of the elastomer. The spatial distribution of the magnetic particles in the matrix can be random (isotropic) or structured (anisotropic). A random distribution is obtained by mixing the ferrous particles throughout the matrix. However, a strong magnetic field must be applied during the composite synthesis to obtain an anisotropic structure. The magnetic field causes the magnetic particles to align towards the magnetic field. Subsequent curing results in immobilised, ordered, chain-like magnetic particle structures. The literature is not consistent for the terminology of orientation of magnetic particles in MRE, and sometimes isotropic MRE is excluded from the category [5–7]. In this chapter, randomly distributed MRE is referred to as 'isotropic MRE' and structured MRE is referred to as 'anisotropic MRE' because both materials respond to an external magnetic field.

Filler loadings are typically 10–70 vol% iron, and the MR effect becomes significant around a filler load of 30 vol%. These high loads of filler in MRE are required to gain a significant MR effect, but may enhance oxidation of the matrix. Moreover, at too high-filler concentrations, the MR effect decreases significantly with respect to filler stiffening. MRE can show non-zero damping and stiffness in the absence of

https://doi.org/10.1515/9783110639018-005

a magnetic field due to interparticle magnetic attraction. The Young's modulus and shear modulus increase significantly if subjected to a magnetic field. This process is reversible for isotropic and anisotropic MRE [8, 9].

Besides the MR effect, MRE also show actuation in response to a magnetic field as a result of the alignment of the embedded magnetisable particles. The attraction or repulsion of these magnetic particles results in an increase or decrease in MRE dimensions. This phenomenon is called 'magnetostriction'. The MR effect as well as magnetostriction of MRE show rapid responses to changes in the magnetic field (i.e., response time <10 ms) [4, 9, 10]. Because of these phenomena, MRE have high potential for various applications and have, therefore, garnered increased interest in the field of smart material research. Suitable applications for MRE are found in automotive, aerospace and civil engineering [4, 11]. This chapter focuses on the current developments of MRE by means of variable matrices, their material properties and improvement for their use in new applications.

5.2 Mechanical properties

Whether MRE are aligned or not has a significant role in their mechanical behaviour. The chain-like structures of aligned magnetisable particles influence the strength of the material drastically depending on coordination of the alignment relative to the applied force. Figure 5.1 illustrates the effect of particle alignment and orientation on the stress–strain behaviour of silicone-based MRE (type K-70A), with 27 vol% carbonyl iron (average particle size of 3.8 μm) [12]. In this figure, 'transversely aligned' denotes applied stress perpendicular to the aligned particles and 'longitudinally aligned' denotes applied stress parallel to the aligned particles. The addition of carbonyl iron particles to the matrix stiffens the MRE significantly because an increase of ≈85% is observed in nominal tensile stress at a nominal strain of 0.1. This increase in stiffness clearly demonstrates the reinforcing ability of iron particles in silicon-based MRE. Thus, favourable matrix–filler interactions take place (van der Waals forces), which result in higher moduli. Anisotropic specimens show a higher moduli and initial stiffness compared with the isotropic MRE. The alignment of the magnetisable particles clearly results in favoured filler–filler interactions. In comparison with the isotropic- dispersed MRE, the nominal tensile stress at a ratio of 0.1 nominal strain increases at approximately 25 and 50% for longitudinally and transversely aligned specimens, respectively. Noteworthy is the difference between the transversely and longitudinally aligned MRE. It is expected that longitudinally aligned carbonyl iron particles would result in stiffer material, but the opposite is observed. This phenomenon may be attributed to the dimensions of the specimen (60 × 10 × 4 mm). Applying a magnetic field of lower field strength (130 *versus* 180 mT) over a significantly smaller distance (a factor 6) results in a higher degree of alignment.

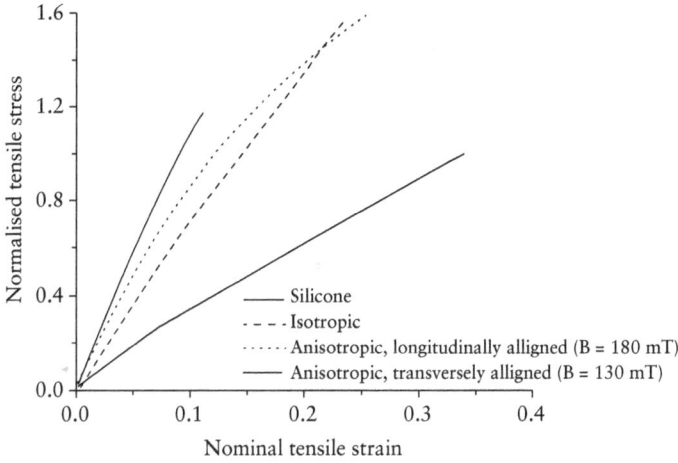

Figure 5.1: Stress–strain curves (normalised for silicone rubber) of unfilled, isotropic-filled and anisotropic-filled silicone rubber (27 vol% carbonyl iron). Adapted from M. Farshad and A. Benine, *Polymer Testing*, 2004, **23**, 3, 347 [12].

The elastic compression moduli of MRE can be determined by unidirectional compression tests. The Neo-Hookean law of rubber elasticity (Equation 5.1) can be used to calculate moduli from obtained datasets presenting non-linear stress–strain relationships, whereas non-ideal systems can be described by Mooney–Rivlin models [13]. In Equation 5.1, 'σ_n', 'G' and 'λ' represent the nominal stress, modulus and deformation ratio (length over the length at zero stress), respectively. Three main configurations can be setup for compression tests of MRE, compression of an: isotropic MRE; anisotropic MRE parallel to the particle alignment; anisotropic MRE perpendicular to the alignment.

$$\sigma_n = G\left(\lambda - \lambda^{-2}\right) \tag{5.1}$$

From a theoretical viewpoint, two responses to compression of an MRE can be observed. For the first type of response, increasing the amount of magnetisable particles does not influence the modulus significantly as long as enough free volume is available for the fillers [i.e., compression does not result in filler arrangement near the critical- filler concentration (CFC)]. This is the case for compression of MRE with isotropic-dispersed magnetisable particles and for compression of MRE with anisotropic-dispersed particles where the direction of compression is perpendicular to the particle alignment. Figure 5.2A and B show this type of response for polydimethylsiloxane (PDMS) matrices filled with carbonyl iron. In both of these figures, it can clearly be seen that an increase in filler content does not result in a significant increase of the moduli. For the second type of response, an increase in the amount of magnetisable particles influences the modulus significantly. This response occurs if

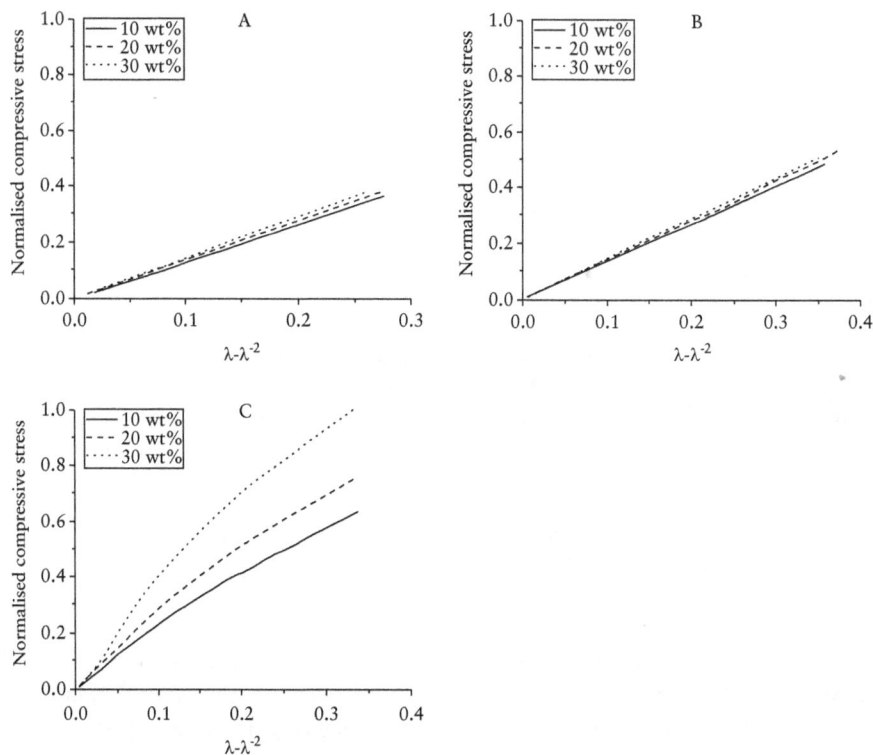

Figure 5.2: Compression measurements of PDMS matrices filled with carbonyl iron (normalised for the maximum compressive stress of 30 wt% anisotropic aligned carbonyl iron with parallel compression) with A) compression of isotropic MRE; B) anisotropic MRE with particle alignment perpendicular to the compression direction; and C) anisotropic MRE with particle alignment parallel to the compression direction. Adapted from Z. Varga, G. Filipcsei and M. Zrinyi, *Polymer*, 2005, **46**, 18, 7779 [13].

the free volume is reduced and local-filler concentrations are over the CFC (i.e., increased filler content results in higher particle-chain density). This is the case for the compression of MRE with anisotropic-dispersed magnetisable particles in which the direction of compression is parallel to the particle alignment, and is clearly shown in Figure 5.2C. In this case, increasing the content of carbonyl iron increases the modulus significantly. This is because under compression, interparticle distances are reduced, which results in increased energy input. Higher crosslink densities yield a more pronounced effect, which may be ascribed to a combination of having a tougher matrix and more tightly locked magnetisable particles.

MRE are generally filled with iron particles, so the oxidative stability of matrices might be influenced by the filler material. This effect is due to the coverage of the iron particle surface by iron oxide [5]. Therefore, relatively high amounts of oxygen can be present in the matrix, which results in partial oxidation. Oxidative resistance of the matrix is generally studied by chemiluminescence and thermal ageing. It has been shown for NR matrices that the incorporation of iron particles (<60 µm) results in significant oxidation of the matrix (Figure 5.3). As can be seen clearly in this figure, the time required to reach a maximum reaction rate decreases with increasing iron content, which indicates lower oxidative resistance of the MRE [5]. The decrease in intensity is ascribed to the emission-shielding ability of iron particles. Moreover, thermal ageing tests have underpinned the conclusions made by chemiluminescence experiments. Ageing of iron particle-incorporated NR resulted in mechanical properties, which strongly indicated the increased oxidation rate of the matrix [5]. Thus, the incorporation of antioxidants into MRE is important for stability, consistency and long-term performance.

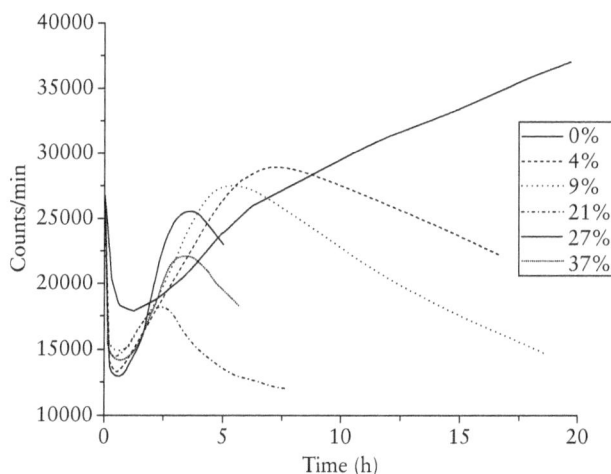

Figure 5.3: Chemiluminescence at 120 °C in iron-filled NR matrices with varying wt%. Adapted from M. Lokander, T. Reitberger and B. Stenberg, *Polymer Degradation and Stability*, 2004, **86**, 3, 467 [5].

5.3 Behaviour in a magnetic field

The MR effect in MRE is significantly dependent on the load of magnetisable particles in the matrix. Dynamic testing of isotropic carbonyl iron-filled silicone in magnetic fields shows a clear effect of the filler content on the storage moduli (G') (Table 5.1). G' increases from ≈25% for a filler content of 10 vol% to >100% for a filler content of 20 vol%, both for the magnetic field strengths of 20 and 600 mT. Obviously, pure silicone shows no response to magnetic fields. For loss moduli (G''), similar trends have been observed [14]. The MR effect becomes even more pronounced if magnetisable particles are aligned. At a filler content of 20 vol%, G'

Table 5.1: Approximate storage moduli of shear stresses (f = 10 Hz) of silicone matrices with carbonyl iron filler with isotropic dispersion in magnetic fields (normalised for G' 20 vol% iron in a magnetic field of 600 mT).

Magnetic field (mT)	Vol% iron	Storage modulus G' (kPa)	
	0	10	20
20	0.36	0.47	0.49
200	0.36	0.51	0.58
400	0.36	0.56	0.80
600	0.36	0.58	1.00

Adapted from H. Boese, *International Journal of Modern Physics B*, 2007, **21**, 28–29, 4790 [14].

increases by a factor of almost 12 for anisotropic silicone filled with carbonyl iron at a magnetic field strength of 600 mT [14].

Compression of MRE in a magnetic field can be undertaken in seven experimental configurations in which the variables are the presence of magnetisable particle alignment, direction of force, and the direction of the magnetic field. An overview of these possible configurations is shown in Figure 5.4.

The energy-density behaviour of MRE in uniform magnetic fields can be described as a function of magnetic and elastic energy contributions (Equation 5.2) [15]. In this

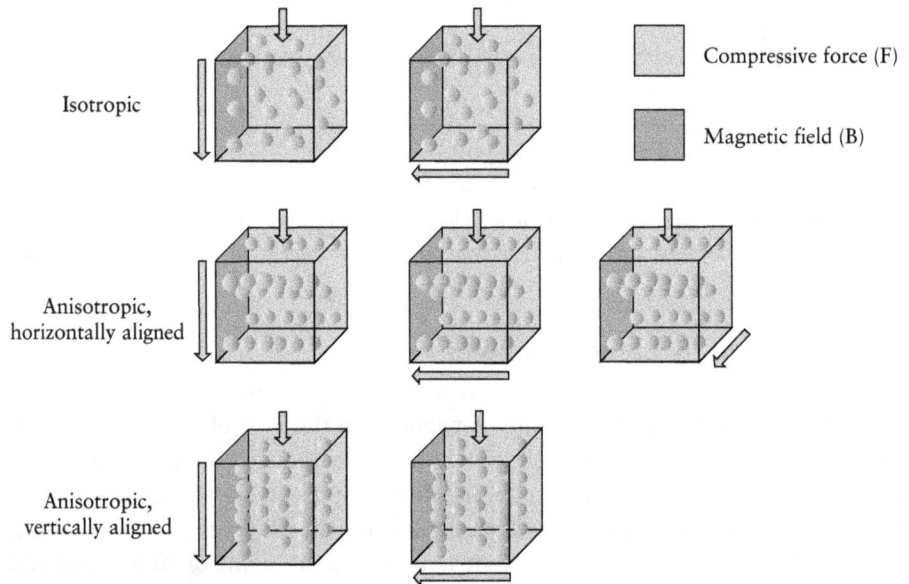

Figure 5.4: Possible experimental configurations for compressing MRE in the presence of a magnetic field. Adapted from Z. Varga, G. Filipcsei and M. Zrinyi, *Polymer*, 2006, **47**, 1, 227 [15].

equation, the elastic contributions 'W_{el}' are dependent on the elastic modulus 'G', where the deformation ratio 'λ_x' (which is parallel to the compressive force) is the variable parameter. The magnetic contributions are dependent on the magnetic permeability of the vacuum 'μ_0' and the magnetic susceptibility of the MRE, where the magnetic field strength 'H_{eff}' is the variable parameter. Thus, as described by Equation 5.2, the stiffness of the MRE is dependent on the intrinsic properties as well as the strength of the magnetic field:

$$W\left(\lambda_x, H_{eff}\right) = W_{el}(\lambda_x) + W_M\left(H_{eff}\right) \tag{5.2}$$

Isotropic PDMS matrices filled with carbonyl iron show a slight increase in elastic modulus in a uniform magnetic field. The magnetic field is perpendicular to the applied force, so the MR effect is slightly more pronounced in comparison with a parallel magnetic field. This observation might be attributed to interparticle attractive forces perpendicular to the compressive force. However, elastic moduli do not increase more than ≈10% with a magnetic field strength of 100 mT in comparison with the zero field modulus. Moreover, the stiffening effect of iron particles is significant with increasing filler content. Compression of anisotropic PDMS matrices filled with carbonyl iron also show a significant MR effect when the applied magnetic field is parallel to the particle alignment. If the magnetic field is parallel to the particle alignment and a compressive force is applied perpendicular to the particle alignment (Figure 5.4), an increase in the elastic modulus of 15–20% is observed when the absolute values of moduli increase with increasing filler content. For the other two scenarios, the MR effect is less pronounced. The appliance of a compressive force parallel to the particle alignment shows similar behaviour. However, zero field moduli are significantly higher, especially in comparison with compression perpendicular to the particle alignment. Moreover, the MR effect increases the elastic modulus >50% with a filler load of 30 wt% if the magnetic field is parallel to the particle alignment [15]. Thus, the orientation of the magnetic field and particle alignment with respect to the compressive force are variables that have a major impact on the MR effect.

The influence of concentration of magnetic-filler content has been investigated for PB-PU matrices (1:1 ratio) filled with carbonyl iron (particle size 3–7 µm) [2]. Compression of the MRE is shown in Figure 5.5. It is clear that the optimum filler concentration is 60 wt% carbonyl iron because, for that sample, the MR effect is significantly stronger than for the samples with 50 and 70 wt% of carbonyl iron. This optimum is ascribed to interparticle free space, or otherwise specified as the CFC. In this case, the CFC is exceeded if >60 wt% of carbonyl iron is incorporated, whereby the effect of alignment is deteriorated by excess magnetic particles by means of decreasing anisotropy.

Fabrication conditions (i.e., by means of temperature) of unsaturated MRE influence the mechanical performance of the material [16]. It has been shown that the

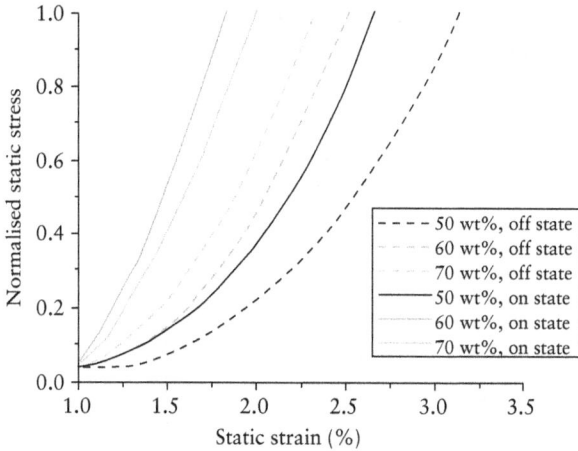

Figure 5.5: Stress–strain curves of PB–PU with aligned carbonyl iron (normalised for a static stress of 35 kPa). Adapted from A. Fuchs, Q. Zhang, J. Elkins, F. Gordaninejad and C. Evrensel, *Journal of Applied Polymer Science*, **105**, 5, 2497 [2].

temperature during the alignment of magnetic particles is an important factor for aligned MRE [16]. For NR with carbonyl iron (60 wt%), it has been found that the optimal MR effect is observed for a configuration temperature of 80 °C (Figure 5.6) with a magnetic field strength of 1 T [16]. This observation is ascribed to the nature of the

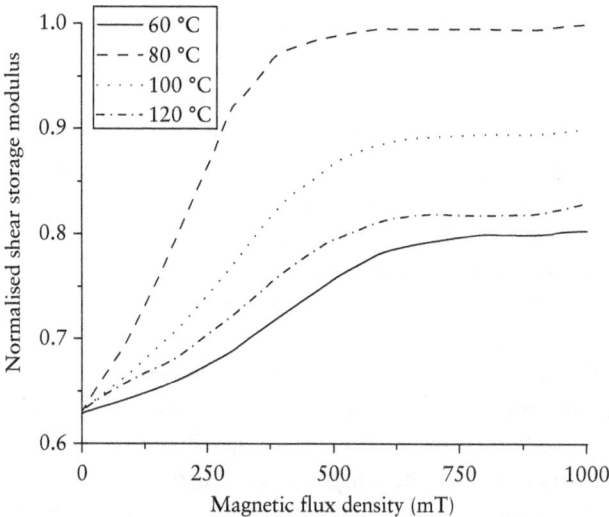

Figure 5.6: Shear modulus of NR with 60 wt% carbonyl iron aligned at various temperatures (normalised for the modulus at 80 °C with a magnetic flux density of 1 T). Adapted from L. Chen, X. Gong, W. Jiang, J. Yao, H. Deng and W. Li, *Journal of Materials Science*, 2007, **42**, 14, 5483 [16].

matrix. With the first increment in alignment temperature, the viscosity decreases, thereby facilitating improved migration of magnetic particles into the magnetic field lines. With the second and third temperature increment (from 80 °C to higher temperatures), crosslinking occurs due to the unsaturated nature of the matrix. Consequently, the viscosity increases, which results in worse migration and alignment of the magnetic particles. Conversely, for saturated matrices, no optimal configuration temperature is expected because the optimal degree of alignment will be reached at a certain temperature. Beyond this temperature no further improvement in alignment is possible.

Films of PDMS filled with fluorinated carbonyl iron (pristine carbonyl iron of 6–9 µm) show controllable surface morphology whereby the surface hydrophobicity can be tuned and superhydrophobic material can be made reversibly [17]. The fluorination of the carbonyl iron is done to increase the hydrophobic character of the particles. Variation of the particle concentration and magnetic field strength result in remarkable changes in surface structure (i.e., magnetostriction) whereby an increase in both variables results in broader and higher rugged surfaces. Hence, the variation in surface roughness results in tuneable wettability of the films. It has been shown that increasing the amount of incorporated fluorinated carbonyl iron from 10 to 70 wt% causes the contact angle of water to increase from ≈98 to 101° up to 96 to 163°, respectively [17]. Moreover, wetting angles do not vary significantly over several cycles of magnetic field application.

5.4 Electrical properties in magnetic fields

The change in permittivity of a material in a magnetic field is called the 'magnetodielectric effect'. The permittivity in MRE is strongly dependent on the type of magnetic filler and is, therefore, dependent on coercivity and saturation magnetisation. As expected, size and quantity also have a significant effect [18]. Magnetically hard fillers, by means of coercivity, show higher magnetodielectric effects than magnetically soft fillers. Hence, the permittivity of a medium can be tuned by orientation of the magnetic field with respect to the capacitor.

Not only the permittivity of the MR material but also the electrical resistance changes if a magnetic field or external force is applied. However, stable conductive behaviour is generally reached after longer periods of time [19]. The issue of slow conduction stabilisation can be overcome by the construction of a magnetoresistor (MRE hybrid), which consists of a resistive inner layer (silicone oil with dispersed graphene nanoparticles) coupled to two magnetoactive elements [silicon rubber (SR) with dispersed carbonyl iron, diameter 4.5–5.4 µm]. The application of a magnetic field and compression changes the resistivity of the magnetoresistor, which becomes stable in <10 s [19]. Due to the response of resistivity to a magnetic field and compression, several potential applications are possible in the construction industry. As a consequence of both responses, the magnetic field intensity can be coupled to compression pressure, which is displayed in Figure 5.7.

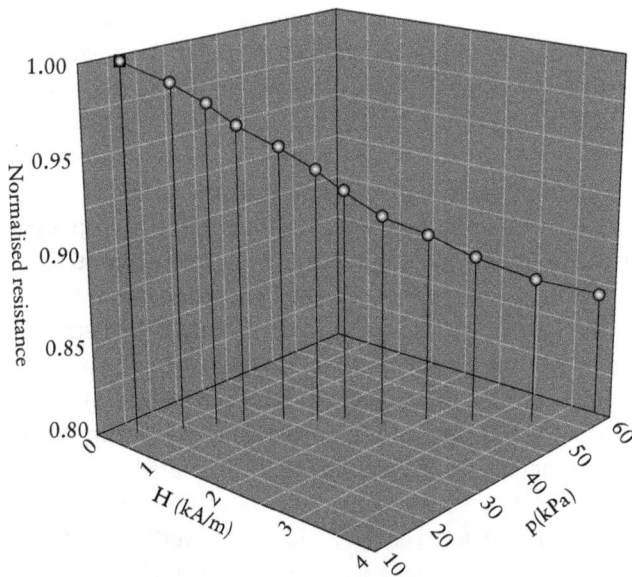

Figure 5.7: Coupling of magnetic field intensity and compression to the resistance of a silicone-based MRE hybrid magnetoresistor (normalised for the resistance at 0.3 kA/m and 14 kPa). Adapted from I. Bica, E.M. Anitas, M. Bunoiu, B. Vatzulik and I. Juganaru, *Journal of Industrial and Engineering Chemistry*, 2014, **20**, 6, 3994 [19].

Electrical resistance of ternary MRE systems can be tuned by varying the ratio of polymer:polymer. For blends of SR/polystyrene (PS) with 75 wt% carbonyl iron, it was found that the volume resistivity decreases in the first stage of PS addition to the MRE (Table 5.2). However, at a ratio of 4:1 SR:PS, the volume resistivity increases significantly. The same trend is visible for the absolute MR effect in a magnetic field of 600 mT [20]. This peculiar trend has not been reported previously. One proposed mechanism for this observation is that the iron particles are preferably located in the SR matrix and both matrices are incompatible with each other by means of agglomeration of PS. The incorporation of PS ratios of 1:19 and 1:9 results in irregular agglomerates that disperse uniformly throughout the continuous SR phase. At a SR:PS ratio of 4:1, spherical PS agglomerates are observed [20]. According to these observations, it might be proposed that for relatively low amounts of incorporated PS, iron particles are located closer together, which results in higher conductivity, but no significant anisotropy is present to increase the MR effect. At a SR:PS ratio of 4:1, PS agglomerates influence the dispersion of carbonyl iron particles sufficiently to create local

Table 5.2: Influence of PS fraction in SR on the volume resistivity and MR effect with 75 wt% carbonyl iron (both normalised at a SR:PS ratio of 1:0).

Ratio SR:PS	Normalised resistivity	Normalised MR effect at an applied magnetic field of 600 mT
1:0	1.00	1.00
19:1	0.03	0.64
9:1	0.10	0.57
4:1	0.19	1.71

Adapted from Y. Wang, Y. Hu, X. Gong, W. Jiang, P. Zhang and Z. Chen, *Journal of Applied Polymer Science*, 2007, **103**, 5, 3143 [20].

anisotropy, which results in a higher MR effect. However, these PS agglomerates cause the volume resistivity to increase.

5.5 Applications

MRE are often used in vibration absorbers because of the MR effect. In contrast with classic dynamic vibration absorbers, the stiffness of MRE can be tuned and hence so can the resonance frequency. Hence, MRE can be designed to function as adaptive tuned vibration absorbers (ATVA). The laminar configuration of 19 iron plates (thickness of 1 mm) between 20 SR sheets (thickness of 2 mm) with 20 wt% isotropic-dispersed iron particles and 10 wt% silicone oil results in a prototype ATVA suitable for seismic mitigation [21]. Change of current through the electromagnet (i.e., applying a magnetic field, with the field strength as a function of current is determined experimentally) in the prototype bearing results in a shift of peak frequency of the ATVA (Figure 5.8). The acceleration transmissibility is the ratio of the force transmitted through the bearing to the exiting force. As can be seen clearly, the peak frequency can be shifted from 10 to 20 Hz over a current range of 0 to 5 A. Due to the low frequency range of tuneable peak frequencies, the prototype bearing is highly suitable for mitigation of seismic activity (frequencies of ≤100 Hz).

Widening the band of the natural frequency results in migration over a broader range. An ATVA based on laminar sheets of steel and MRE sheets of 70 wt% carbonyl iron, 15 wt% silicone rubber and 15 wt% silicone oil (sheet thickness of 1 mm) shows two natural frequencies (i.e., translational and torsional) [22]. Increasing the current of the electromagnet shifts the natural frequency of the ATVA for both natural frequencies. In contrast with the ATVA discussed above, the presence of two natural frequencies makes this specific ATVA suitable for applications that require broad effective bandwidth damping and multi-frequency damping.

MRE can change their shape in a magnetic field, so they can be applied as actuating material. Therefore, MRE can be developed into flow-control valves. Fabrication

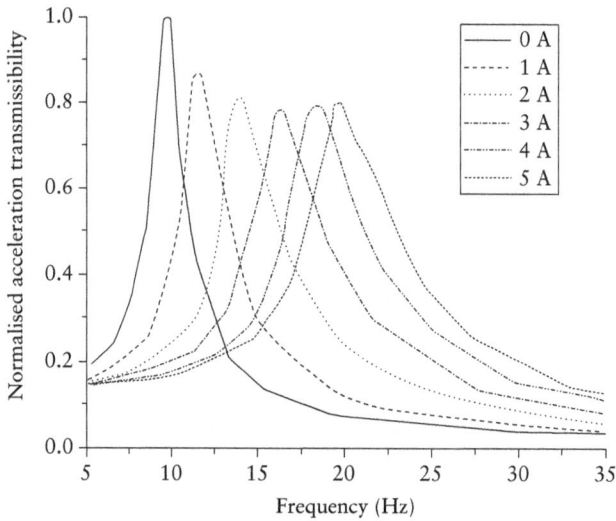

Figure 5.8: Transmissibility of silicone-based ATVA *versus* excitation frequency (normalised for the maximum acceleration transmissibility with 0 A). An increase in the current through the electromagnet results in a shift of peak frequency. Adapted from Z. Xing, M. Yu, J. Fu, Y. Wang and L. Zhao, *Journal of Intelligent Material Systems and Structures*, 2015, **26**, 14, 1818 [21].

of ring-shaped PDMS with varying crosslink densities filled with 30 wt% carbonyl iron results in successful magnetoactive flow-control valves [23]. Shore hardness, particle arrangement and the shape of the yoke are important parameters to take into account for the design of the valve. It has been shown that an increase in hardness requires a higher current and, therefore, a stronger magnetic field to achieve equal air flow rates with respect to softer MR material (shore hardness of 55–70), especially in a low-flow rate regime. This behaviour is in agreement with the material properties because an increase of stiffness requires a stronger magnetic field for deformation. Thus, the sensitivity of the magnetoactive valve to a magnetic field can be tuned by varying the crosslink density. Arrangement of magnetic particles is important to the flow behaviour as a function of the current applied to the electromagnet. The isotropic magnetoactive valve shows a stronger linear relationship between current and flow rate with respect to a radially aligned anisotropic magnetoactive valve.

The fabrication of MRE brushes results in a smart structure that makes use of the MR effect as well as buckling instability. The critical load of a column can be calculated by Euler's critical-load formula (Equation 5.3):

$$P_{CR} = \frac{\pi^2 EI}{(KL)^2}$$

(5.3)

In Equation 5.3, 'P_{CR}' is the critical load (N), 'E' is the elastic modulus (Pa) and 'I' is the second moment of inertia (m^4). MRE hybrid materials filled with conductive particles can be designed for the application of magnetic field sensors. The working principle behind this application is based on magnetostriction. An elastomeric material is filled with homogeneously-dispersed ferromagnetic and conductive particles and subsequently crosslinked, so all particles are locked into the matrix. Application of a voltage to the MRE hybrid results in current flow though the material. Subsequently, once a magnetic field is applied perpendicular to the direction of the electrical current, the MRE hybrid change shapes. As a consequence, the interparticle distance of conductive particles and the resistance decreases. It has been shown for a MRE hybrid based on 40 wt% SR, 15 wt% silicone, 20 wt% carbonyl iron (d_{avg} of 0.612 µm) and 20 wt% graphite that the current and output voltage increases with increasing magnetic field strength [24]. By increasing the applied voltage to the magnetic field sensor, an absolute increase in the current and output voltage is observed. However, the increase in current output voltage is not linear relative to a linear increase of applied current. One possible explanation for this observation might be the influence of the Joule effect, though thermal analysis is required to underpin this theory.

References

1. J. Rabinow, *Transactions of the American Institute of Electrical Engineers*, 1948, **67**, 2, 1308.
2. A. Fuchs, Q. Zhang, J. Elkins, F. Gordaninejad and C. Evrensel, *Journal of Applied Polymer Science*, 2007, **105**, 5, 2497.
3. J. Ginder, S. Clark, W. Schlotter and M. Nichols, *International Journal of Modern Physics B*, 2002, **16**, 17–18, 2412.
4. Y. Wang, Y. Hu, Y. Wang, H. Deng, X. Gong, P. Zhang, W. Jiang and Z. Chen, *Polymer Engineering and Science*, 2006, **46**, 3, 264.
5. M. Lokander, T. Reitberger and B. Stenberg, *Polymer Degradation and Stability*, 2004, **86**, 3, 467.
6. N. Kchit and G. Bossis, *Journal of Physics: Condensed Matter*, 2008, **20**, 20, 204136.
7. Y. Yu, Y. Li and J. Li, *Journal of Intelligent Material Systems and Structures*, 2015, **26**, 18, 2446.
8. Y.F. Zhou, S. Jerrams, A. Betts and L. Chen in *Proceedings of the 8th European Conference on Constitutive Models for Rubbers (ECCMR)*, 25–27th June, San Sebastian, Spain, 2013, p.683.
9. Y. ZuGuang, N. YiQing and M. Sajjadi, *Science China Technological Sciences*, 2013, **56**, 4, 878.
10. G. Du and X. Chen, *Measurement*, 2012, **45**, 1, 54.
11. M.P. Vasudevan, P.M. Sudeep, I.A. Al-Omari, P. Kurian, P.M. Ajayan, T.N. Narayanan and M.R. Anantharaman, *Bulletin of Materials Science*, 2015, **38**, 3, 689.
12. M. Farshad and A. Benine, *Polymer Testing*, 2004, **23**, 3, 347.
13. Z. Varga, G. Filipcsei and M. Zrinyi, *Polymer*, 2005, **46**, 18, 7779.
14. H. Boese, *International Journal of Modern Physics B*, 2007, **21**, 28–29, 4790.
15. Z. Varga, G. Filipcsei and M. Zrinyi, *Polymer*, 2006, **47**, 1, 227.

16. L. Chen, X. Gong, W. Jiang, J. Yao, H. Deng and W. Li, *Journal of Materials Science*, 2007, **42**, 14, 5483.
17. S. Lee, C. Yim, W. Kim and S. Jeon, *ACS Applied Materials & Interfaces*, 2015, **7**, 35, 19853.
18. A.S. Semisalova, N.S. Perov, G.V. Stepanov, E.Y. Kramarenko and A.R. Khokhlov, *Soft Matter*, 2013, **9**, 47, 11318.
19. I. Bica, E.M. Anitas, M. Bunoiu, B. Vatzulik and I. Juganaru, *Journal of Industrial and Engineering Chemistry*, 2014, **20**, 6, 3994.
20. Y. Wang, Y. Hu, X. Gong, W. Jiang, P. Zhang and Z. Chen, *Journal of Applied Polymer Science*, 2007, **103**, 5, 3143.
21. Z. Xing, M. Yu, J. Fu, Y. Wang and L. Zhao, *Journal of Intelligent Material Systems and Structures*, 2015, **26**, 14, 1818.
22. S. Sun, J. Yang, W. Li, H. Deng, H. Du, G. Alici and T. Yan, *Smart Materials and Structures*, 2016, **25**, 5, 055035.
23. H. Boese, R. Rabindranath and J. Ehrlich, *Journal of Intelligent Material Systems and Structures*, 2012, **23**, 9, 989.
24. I. Bica, *Journal of Industrial and Engineering Chemistry*, 2011, **17**, 1, 83.

6 Dielectric elastomers

6.1 Introduction and basic principles

Dielectric elastomers (DE) are smart materials that belong to the group of electroactive polymers (i.e., polymers that show a significant change in shape under electrical stimuli). DE are constructed by 'sandwiching' an elastomeric membrane between two stretchable electrodes that are in direct contact with the elastomer membrane. Typically, acrylic- and silicone-based polymers are used as elastomeric layers [1], whereas carbon grease, carbon nanotube coatings, graphene sheets and metallic nanoclusters are used as electrodes [2]. The application of a voltage to the electrodes, which act as capacitors, creates an electric field. Actuation behaviour can then be observed as a response to the application of this electric field when coulombic forces between the two electrodes causes them to attract each other and, in turn, the DE squeezes between the electrodes and expands in area (Figure 6.1). Voltages ranging from 0.5 to 10 kV are commonly applied to reach satisfactory actuation behaviour (up to areal strains of 1,000%) [3]. Examples of applications of DE actuators are robotics, speakers, braille displays, pumping devices and motors [1, 4].

The pressure induced by the electrostatic interaction (i.e., Maxwell stress) is described by Equation 6.1, where 'p' is the Maxwell stress, 'A' the area of the DE, 'U' the electrostatic energy, and 'z' the thickness of the DE [5]:

$$P = A^{-1}\left(-\frac{dU}{dz}\right)$$
(6.1)

The electrostatic energy stored in a capacitor is expressed according to Equation 6.2, where the capacitance 'C' is defined by Equation 6.3 ('ε_0' is the permittivity of vacuum and 'ε_r' is the relative permittivity of the medium between the two electrodes, i.e., the elastomer), and 'V' is the applied voltage.

$$U = \frac{1}{2}CV^2$$
(6.2)

$$C = \varepsilon_0\varepsilon_T\frac{A}{Z}$$
(6.3)

Applying the equation for the capacitance to the equation for the Maxwell stress to incompressible DE (i.e., with a conserved volume) yields Equation 6.4, where 'E' is the electric field strength [5]:

$$P = \varepsilon_0\varepsilon_T\left(\frac{V}{Z}\right)^2 = \varepsilon_0\varepsilon_T E^2$$
(6.4)

Finally, the compressive strain in a DE is expressed by Equation 6.5, which is obtained by the application of Hooke's law to Equation 6.4, where 'S' is the

https://doi.org/10.1515/9783110639018-006

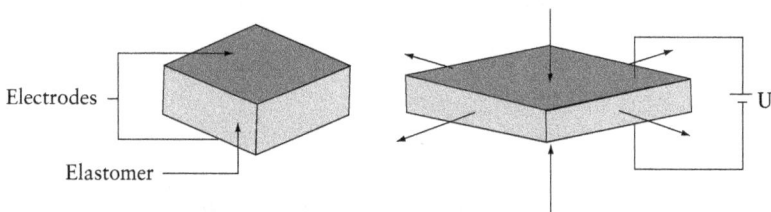

Figure 6.1: Operating principle of actuation by DE. Application of a voltage results in stretching of the elastomer in a parallel direction and compression in a perpendicular direction of the electrodes. Adapted and redrawn from R. Pelrine, R. Kornbluh and G. Kofod, *Advanced Materials*, 2000, **12**, 16, 1223 [1].

compressive strain and 'Y' is the Young's modulus. In Equation 6.5, the Young's modulus is assumed to be constant and the compression of DE shows a Poisson coefficient of 0.5 (i.e., compression along the z-axis results in elongation of the x and y axes with factor of 0.5) [5].

$$S = -\frac{P}{Y} = -\varepsilon_0 \varepsilon_T \frac{E^2}{Y} \tag{6.5}$$

When considering Equations 6.4 and 6.5, it can be concluded that the actuation of a DE is determined by the permittivity of the elastomer, the applied voltage, the distance between the two electrodes and the Young's modulus of the elastomer. Hence, the elastomer must exhibit a low modulus and high permittivity to obtain large deformation within a DE. A low Young's modulus can be achieved by, for example, limiting the crosslink density of the elastomeric matrix or by the introduction of bulky side groups on the macromolecular backbone. Conversely, an increase in permittivity can be obtained by the introduction of polarisable functionalities or permanent dipoles. Moreover, the thickness of the DE is important, as evidenced in Equation 6.5. Consequently, DE are generally pre-stretched before an electrical field is applied, resulting in a decrease of the thickness of the membrane and, in turn, in a higher electrical field strength through the DE.

DE generally display two types of flaws: electromechanical instability and electrical breakdown [3]. Electromechanical instability is shown by positive feedback that is typically described by amplified deformation. DE are squeezed upon actuation, so the electrical field strength increases in response to a decrease in electrode distance, and premature electrical breakdown may be observed as a result [3]. Electrical breakdown occurs at the electric field strength at which the elastomeric membrane becomes conductive. As a consequence, electrostatic attraction is lost and the actuation effect vanishes. In this context, it is noteworthy that DE should exhibit low Young's moduli and high dielectric constants. Remarkably, these two requirements specifically induce the two previously mentioned flaws in DE. This contradiction clearly demonstrates the difficulty of developing high-end-performance DE actuators. This chapter focuses on recent developments in systems for DE using tailored elastomers (i.e., improved matrices by chemical functionalisation and physical incorporation of additives in DE).

6.2 Tailored elastomers

The development of copolyesters based on five monomers (1,4-butanediol, 1,3-propanediol, sebacic acid, succinic acid and itaconic acid) results in a DE that displays sufficient actuation strain (11.9%, $E = 15.6$ kV/mm) without the need to pre-strain the material [6]. The crosslink density of this copolymer can be tuned by varying the reaction conditions and the feed ratios of the monomers [6]. It has been found that, for this material, increasing the crosslink density results in an increased dielectric constant and an increased Young's modulus (Figure 6.2). The Young's modulus of this material is in the same order of magnitude of common bench-mark DE materials, such as acrylic rubber VHB^{TM} 4910, whereas the dielectric constant of the copolyester DE is even higher than that of VHB 4910. This observation is ascribed to the high amount of polarisable groups in the copolyester that favour the interactions among the different macromolecules [6].

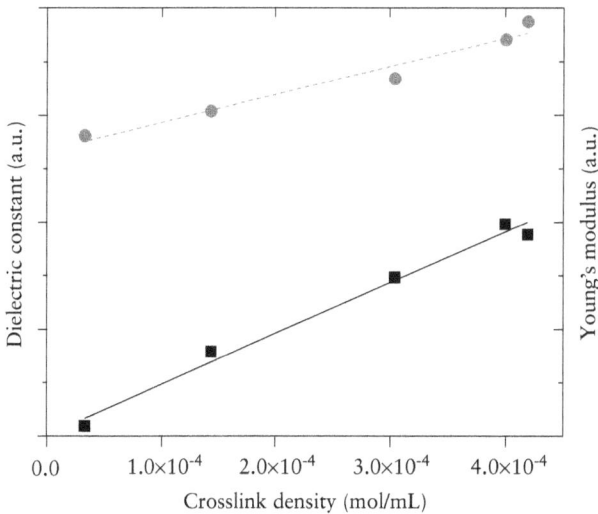

Figure 6.2: Dielectric constant [alternating current (AC) of 1 kHz shown in grey] and the Young's modulus (black) as a function of crosslinking density. Adapted and redrawn from D. Yang, M. Tian, H. Kang, Y. Dong, H. Liu, Y. Yu and L. Zhang, *Material Letters*, 2012, **76**, 1, 229 [6].

When taking the parameters mentioned above into consideration, it appears that the selection of the monomer is critical for the parameters that govern the DE actuation for the design of new polymeric materials. According to Equation 6.5, it is of crucial importance how the ratio of the dielectric constant over the modulus progresses with increasing crosslink density. Consequently, it can be deduced that an increase in crosslink density results in strains of higher actuation. From Figure 6.3, it is worth stressing that the ratio of the dielectric constant over the Young's modulus is the highest for materials with the lowest degree of crosslinking ($3.28 \times 10^{-5}\,\mathrm{mol/cm^3}$).

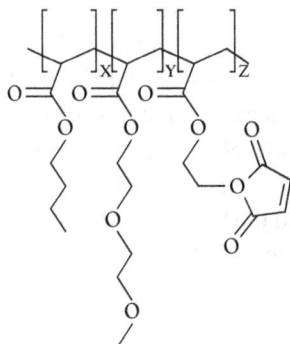

Figure 6.3: Molecular structure of a tailored polyacrylate with DA dienophilic maleimide functional groups.

Moreover, the highest-actuation strain is also observed at the lowest crosslink density. More specifically, an actuation strain of 11.9% is observed for a crosslink density of $3.28 \times 10^{-5}\,\mathrm{mol/cm^3}$ and the actuation strain is more than halved when increasing the crosslink density to $4.19 \times 10^{-4}\,\mathrm{mol/cm^3}$. However, the electrical breakdown strength also increases with an increased in crosslink density.

Tuneable DE systems can be developed by the introduction of reversible crosslinks into the elastomeric membrane. Such reversible crosslinks can be obtained by applying several strategies that have been discussed in Chapter 1, for example, thermo-reversible Diels–Alder (DA) chemistry [7]. The sophisticated synthesis of a maleimide functional polyacrylate results in such an elastomer (Figure 6.3). In this case, the addition of a crosslinker with four functional furan groups yields a modulus-tuneable DE. The intrinsic properties of the resulting material can be tuned by varying the ratios of the monomers X, Y and Z. For example, an increased amount of maleide groups (component Z in Figure 6.3) results in a higher crosslink density, and increasing the amount of 2-(2-ethoxyethoxyl) ethyl acrylate (component Y in Figure 6.3) would theoretically result in an increased relative permittivity due to functional groups that are relatively more polarisable.

Subjecting the exemplified DA crosslinked polyacrylate to different thermal treatments shows that the stiffness of the material is indeed tuneable as function of the temperature of the treatment. For example, a thermal treatment at 130 °C (i.e., de-crosslinking by the retro-DA reaction) results in a decrease in Young's modulus of ≈50%. As expected, the actuation strain increases upon de-crosslinking the maleimide polyacrylate DE. However, the crosslinked polyacrylate is preferential for high-voltage applications because it displays a higher breakdown strength and better electromechanical stability [7].

The functionalisation of polymers with permanent dipoles for polymers that are suitable for DE actuators should theoretically result in increased permittivity (Equation 6.4). Vinyl-terminated polydimethylsiloxane (PDMS) with azidopropyl-functionalised groups is an example of a polymer that can be functionalised with permanent dipoles *via* 'click chemistry' reactions. In this case, 1-ethynyl-4-nitrobenzene (Figure 6.4), which contains a permanent dipole, can be 'clicked' onto the azidopropyl-functionalised vinyl- terminated PDMS *via* a copper-catalysed cycloaddition. Subsequently, crosslinking with a hydride functional siloxane-based crosslinker results in elastomer that contains permanent dipoles [8].

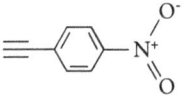

Figure 6.4: Molecular structure of 1-ethynyl-4-nitrobenzene.

It has been found that the spacing between the nitrobenzene functionalities in DE has a significant influence on DE permittivity (Figure 6.5). In this work, the material with 0 wt% nitrobenzene does not refer to the pristine crosslinked PDMS, but instead to azidopropyl-functionalised PDMS. Both types of nitrobenzene-functionalised PDMS DE have an optimum of nitrobenzene content. However, a spacing of 1,200 g/mol between the functional groups results in a shift of this optimum to lower weight percentages and lower relative permittivity with respect to spacing of 580 g/mol. One explanation for this observation might be that local agglomerates of nitrobenzene functionalities are formed more easily at a lower degree of spacing, which may result in highly dielectric fractions in the DE. Unfortunately, the highest electrical breakdown is observed for azidopropyl-functionalised PDMS with spacing of 580 g/mol. Apparently, the trade-off between permittivity and modulus does not result in improved stability for the DE, whereas functionalisation with nitrobenzene results in a significant increase in the dielectric constant [8].

In contrast with the nitrobenzene-functionalised PDMS discussed above, the introduction of chloropropyl functionalities in a PDMS DE results in increased permittivity as well as electrical breakdown [4]. However, this is based on a blend of pristine PDMS with chloropropyl- functionalised PDMS. Nonetheless, this is a remarkable result. The chloropropyl-functionalised PDMS is synthesised using a similar procedure to that described for the nitrobenzene-functionalised PDMS. However, click chemistry cannot be applied because the chloropropyl functionality is incorporated in the polymer in the

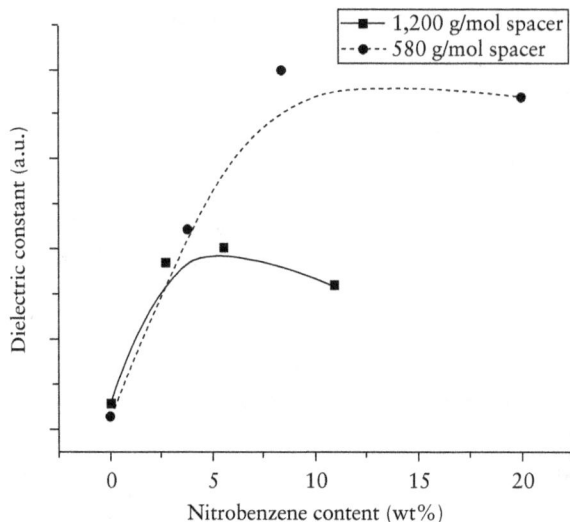

Figure 6.5: Progression of the dielectric constant as function of the nitrobenzene content grafted on PDMS with spacing of 1,200 and 580 g/mol at 1 kHz AC. Adapted and redrawn from F.B. Madsen, L. Yu, A.E. Daugaard, S. Hvilsted and A.L. Skov, *Polymer*, 2014, **55**, 24, 6212 [8].

same manner as that for the azidopropyl precursor of the nitrobenzene- functionalised PDMS. Crosslinking of this material is done using the same methodology, and the spacing between the functionalities is again 580 and 1,200 g/mol, thereby allowing for a fair comparison. The highest electrical breakdown (101.2 V/µm) is observed for the chloropropyl-functionalised PDMS with spacing of 580 g/mol and 75 wt% pristine PDMS. This is an increase in electrical breakdown of ≈25% with respect to pristine PDMS [4]. For comparison, the functionalisation of PDMS with nitrobenzyl and chloropropyl results in increased permittivity in both cases, but electrical breakdown is increased only for the chloropropyl-functionalised PDMS.

In conclusion, the polymer design of DE appears to be highly important, especially with respect to permittivity, the Young's modulus and ultimate tensile strength. Examples of newly designed DE have shown that these dielectric properties can be tuned appropriately by selecting polarisable monomers and controlling the crosslink density.

6.3 Additives in dielectric elastomers

Besides polymer functionalisation, increased dielectric properties of DE can also be achieved by the incorporation of additives in polymer matrices. One advantage of using additives as dielectric enhancers over polymer functionalisation can be that some additives have a plasticising effect. This results in a lower Young's modulus, yielding a more pronounced actuation but a reduction in the electrical breakdown, which is undesirable. Additionally, the syntheses of filled DE are commonly straightforward in comparison with the functionalisation of DE.

The addition of chloropropyl-functionalised silicone oil to PDMS results in significantly increased permittivity while not deteriorating the electrical breakdown (Figure 6.6). An increase in the electrical breakdown is observed only in a regime of low amounts of additive. The electrical breakdown decreases significantly upon the addition of larger amounts of additive. The same is true for the Young's modulus [9]. From a theoretical viewpoint, the actuation strain of PDMS filled with chloropropyl-functionalised silicone oil should, therefore, be significantly higher than that of pristine PDMS. This is the case for PDMS filled with 30-phr chloropropyl-functionalised silicone oil as it displays approximately doubled-actuation strains in comparison with pristine PDMS. Moreover, the addition of 30-phr chloropropyl-functionalised silicone oil to PDMS filled with 30-phr titanium dioxide (a commonly used polarisable filler) results in enhancement of permittivity and electrical breakdown, which both increase by ≈10% [9].

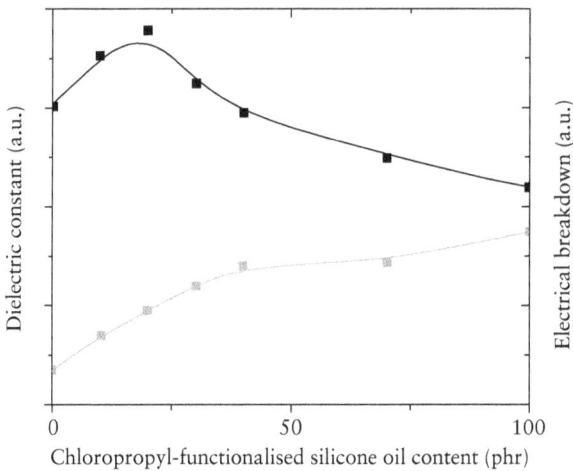

Figure 6.6: Dielectric constant (grey) and electrical breakdown (black) as a function of chloropropyl-functionalised silicone oil content in PDMS (ELASTOCIL® LR 3043/50). Adapted and redrawn from F.B. Madsen, L. Yu, P. Mazurek and A.L. Skov, *Smart Materials and Structures*, 2016, **25**, 7, 075018 [9].

A final example of an additive suitable for dielectric enhancement in DE is glycerol. Figure 6.7 displays the dielectric constant of two types of PDMS matrices: Sylgard® 184 and Powersil® XLR 630 A/B. In this figure, it can clearly be seen that the dielectric constant increases almost linearly with increasing glycerol content for both PDMS types. The addition of a PDMS–polyethylene glycol surfactant does

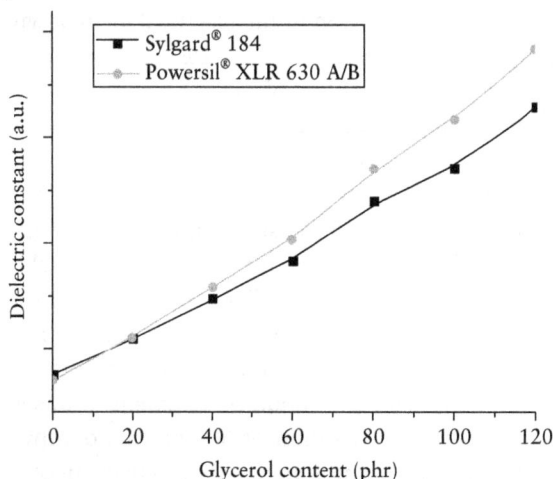

Figure 6.7: Dielectric constant at AC of 1 kHz as a function of glycerol content in two types of PDMS matrices. Adapted and redrawn from P. Mazurek, L. Yu, R. Gerhard, W. Wirges and A.L. Skov, *Journal of Applied Polymer Science*, 2016, **133**, 43, 44153 [10].

not result in a change in permittivity. The conductivity increases with an increasing amount of glycerol in the PDMS matrices as well. Therefore, it might be expected that the electrical breakdown is lowered, which is undesirable. Nonetheless, the addition of glycerol might be considered as a suitable 'green' additive to DE (e.g., in comparison with the chloropropyl-functionalised silicone oil mentioned previously) [10].

References

1. R. Pelrine, R. Kornbluh and G. Kofod, *Advanced Materials*, 2000, **12**, 16, 1223.
2. J.A. Rogers, *Science*, 2013, 341, **6149**, 968.
3. F. Zhu, C. Zhang, J. Qian and W. Chen, *Journal of Zhejiang University: Science A*, 2016, **17**, 1, 1.
4. F.B. Madsen, L. Yu, A.E. Daugaard, S. Hvilsted and A.L. Skov, *RSC Advances*, 2015, **5**, 14, 10254.
5. L.J. Romasanta, M.A. Lopez-Manchado and R. Verdejo, *Progress in Polymer Science*, 2015, **51**, 188.
6. D. Yang, M. Tian, H. Kang, Y. Dong, H. Liu, Y. Yu and L. Zhang, *Materials Letters*, 2012, **76**, 1, 229.
7. W. Hu, Z. Ren, J. Li, E. Askounis, Z. Xie and Q. Pei, *Advanced Functional Materials*, 2015, **25**, 30, 4827.
8. F.B. Madsen, L. Yu, A.E. Daugaard, S. Hvilsted and A.L. Skov, *Polymer*, 2014, **55**, 24, 6212.
9. F.B. Madsen, L. Yu, P. Mazurek and A.L. Skov, *Smart Materials and Structures*, 2016, **25**, 7, 075018.
10. P. Mazurek, L. Yu, R. Gerhard, W. Wirges and A.L. Skov, *Journal of Applied Polymer Science*, 2016, **133**, 43, 44153.

7 Future outlook

'Rubbers are everywhere'. Indeed, life involves the use of rubber products on a daily basis. Every time we drive a car, we rely on more than 100 years of history and development to make sure that the tyres are up to specifications. These specifications determine how the car is made to stop or run at the desired speed, sometimes causing considerable differences in fuel usage. Rubber products are the result of more than 150 years of history and constitute a paradigm of technological advancements in the design of structured and formulated chemical products.

Some of the problems inherent to rubber products remain largely unsolved. The overall issue of recycling, preferably *via* a 'cradle-to-cradle' approach, is a hot topic for researchers and engineers as societal demands in this respect become increasingly pressing. This dichotomy between the advanced technological level of the design and the complete lack of re-workability and recyclability sets very clear and ambitious goals for research at academic and industrial levels.

Against this backdrop, future research will probably have to address the key issue of crosslinking rubbers. The presence of strong, covalent bonds between the polymeric chains is a *conditio sine qua non* for many commercial elastomers to achieve the desired elasticity and high-temperature performance. The same bonds clearly hinder any possibility of recycling and reprocessing of the material. This paradox is inversed in the case of thermoplastic rubbers, where the presence of weak interactions (e.g., hydrogen bonding and van der Waals forces) allows full recyclability, but simultaneously prohibits the application of these materials at relatively high temperatures (above the glass transition temperature).

In the last 30 years, several research teams have published scientific articles on trying to solve the recyclability of vulcanised rubbers and high-temperature applications of thermoplastic rubbers, but significant steps in either direction have not been made. The dated characteristics of the materials and the perceived lack of novelty (in terms of scientific publications) might relegate this research topic to an industrial dimension. Conversely, many insights might be gained when questioning the overall concept of covalent crosslinking. The recent introduction of the concept of reversible networks is a paradigmatic example of a scientifically attractive new insight. Such development often clashes with the conservative attitude of the industrial world that is very reluctant in leaving an efficient and established technology (sulfur vulcanisation or peroxide curing) to start re-optimisation studies for other reaction possibilities. An increase in societal pressure might change this situation and will almost certainly generate renewed interest in these research topics.

In this context, the overall issue of sustainability, in terms also of the bio-based character of materials, might represent an interesting opportunity. This is especially relevant when making allowances for the possible gains (i.e., a reduced carbon

https://doi.org/10.1515/9783110639018-007

footprint). Conversely, other technological advancements, such as the advent of electric cars, could disturb the overall picture because new materials with novel properties (and thus, chemical structure) will be required for different parts of such new cars. Hence, it is not merely a question of changing formulations: the design of new elastomeric products is required as well as the use of novel scientific principles that have yet to be investigated. Therefore, further research on bulk rubber products is heavily dependent on external stimuli and the driving forces of society, industry and academia.

The situation becomes entirely different when dealing with applications for specialty products, as shown in this book. Sensor properties, self-healing, actuators, magneto-rheology, optical properties, and shape memory might drive and steer many future research projects towards the development of novel materials. The relatively small scale of the envisioned specialty applications with respect to the large-scale production of conventional bulk products does not introduce initial constraints on research (e.g., in terms of final product costs, the availability of the mater0069als or their production). In this sense, the role of scientific curiosity and investigation will remain pivotal.

Chemistry, physics and engineering are the basic disciplines required for the design of smart rubbers. In general, the end products do not solely consist of an elastomeric material, but instead are formulated in combination with other polymers and/or unconventional additives. Being typically polymer blends and composites, the properties of the end products are not a mere combination of each component, but a result of the synergy occurring at the interface. The latter can be tuned or manipulated by controlling the phenomena between the components. This interaction is also a function of the chemistry and properties of the functional groups that are present along the polymeric backbone.

We envision research aimed at the specific modification of elastomers in general and at the control over the adhesion at the interface in particular as necessary elements for control of the properties of the final product. Such integrated approaches (from molecules to end products) are pivotal in ensuring industrial applicability because they relate the structure of the product with its properties. Establishment of such structure–property relationships is also crucial in ensuring the design of products by starting with the desired final properties and performance (i.e., the 'market pull' as opposed to the more conventional 'technology push'). The number of possible combinations of rubbers with other polymers and/or fillers is endless, which indicates the possibility of new discoveries in these fields. In this context, research on specialty elastomeric products (as opposed to bulk ones), while still stimulated by external factors, can be also driven by scientific curiosity. In this sense, the corresponding chapters of this book might be considered to be good starting points, but describe findings that pave the way towards many more future developments.

Abbreviations

3D	Three-dimensional
AC	Alternating current
AFD	Astigmatic focal distance
AgNP	Silver nanoparticles
AgNW	Silver nanowires
ALMA	Allyl methacrylate
ATVA	Adaptive tuned vibration absorbers
BBS	*Bis*(benzoxazolyl)stilbene
BDDA	Butanediol diacrylate
BR	Butyl rubber
CB	Carbon blacks
CFC	Critical-filler concentration
CIE	Crystallisation-induced elongation
CNC	Cellulose nanocrystals
CN–OPV	Cyano-functionalised hydroxyl-terminated oligo(p- phenylene vinylene)
CNT	Carbon nanotubes
DA	Diels–Alder
DCJTB	4 -(Dicyanomethylene)-2-t-butyl-6-(1,1,7,7 - tetramethyljulolidyl-9-enyl) -4H-pyran
DCP	Dicumylperoxide
DE	Dielectric elastomers
EDOT	3,4-Ethylenedioxythiophene
EPDM	Ethylene propylene diene rubber
EPM	Ethylene propylene rubber copolymer
GF	Gauge factor
HEMA	Hydroxyethyl methacrylate
HNA	High-aspect ratio nanopillar array
HPU	Hyperbranched polyurethane
IR	Infrared
LC	Liquid crystalline
LDH	Layered double hydroxides
LED	Light-emitting diodes
LNA	Low-aspect ratio nanopillar array
MIC	Melt-induced contraction
MNA	Medium-aspect ratio nanopillar array
MR	Magnetorheological
MRE	Magnetorheological elastomers
MRF	Magnetorheological fluids
MW	Molecular weight
MWCNT	Multi-walled carbon nanotubes
NBR	Acrylonitrile-butadiene rubber
NR	Natural rubber
PANI	Polyaniline
PB	Polybutadiene
PCL	Polycaprolactone
PDMS	Polydimethylsiloxane
PEA	Polyethyl acrylate

https://doi.org/10.1515/9783110639018-008

PEDOT	Poly(3,4-ethylenedioxythiophene)
PGA	Polyglycolic acid
PLLCA	Poly(L-lactide-*co*-ε-caprolactone)
PMMA	Poly(methyl methacrylate)
Ppy	Polypyrrole
PS	Polystyrene
PSS	Polystyrene sulfonate
PU	Polyurethane(s)
rDA	Retro-Diels–Alder
R_f	Retention rate
RGO	Reduced graphene oxide
RI	Refractive index
R_r	Shape-recovery rate
RT	Room temperature
SBR	Styrene-butadiene rubber
SBS	Polystyrene-*b*-polybutadiene-*b*-polystyrene
scCO$_2$	Supercritical carbon dioxide
SD	Standard deviation
SEBS	Polystyrene-*b*-(ethylene-*co*-butylene)-*b*-styrene
SiNC	Silicon nanocrystals
SME	Shape-memory elastomers
SMP	Shape-memory polymers
SR	Silicon rubber
SWCNT	Single-walled carbon nanotubes
T_g	Glass transition temperature
THF	Tetrahydrofuran
TiC	Titanium carbide
TPE	Thermoplastic elastomers
TPU	Thermoplastic polyurethanes
TPV	Thermoplastic vulcanisates
UV	Ultraviolet
UV-Vis	Ultraviolet-visible
VOC	Volatile organic compounds
ZnO	Zinc oxide
ZnS	Zinc sulfide
ΔT	Thermal crosslinking

Index

https://doi.org/10.1515/9783110639018-009